The WOODWORKER'S PROJECT BOOK

The WOODWORKER'S PROJECT BOOK

42 shop-tested plans for easy-to-make gifts, toys & furniture

By the Editors of The Woodworker's Journal

We at Madrigal Publishing have tried to make this book as accurate and correct as possible. Plans, illustrations, photographs, and text have been carefully researched by our in-house staff. However, due to the variability of all local conditions, construction materials, personal skills, etc., Madrigal Publishing assumes no responsibility for any injuries suffered or damages or other losses incurred that result from material presented herein. All instructions and plans should be carefully studied and clearly understood before beginning any construction.

For the sake of clarity, it is sometimes necessary for a photo or illustration to show a power tool without the blade guard in place. However, in actual operation, always use blade guards (or other safety devices) on power tools that are equipped with them.

©1992 by Madrigal Publishing Co. All rights reserved. No part of this book may be reproduced in any form without the permission of the publisher.

Printed in the United States of America.

Library of Congress Cataloging-in-Publication Data:
The Woodworker's project book : 42 shop-tested plans for easy-to-
   make gifts, toys & furniture / by the editors of the Woodworker's
   journal.
      p.   cm.
   ISBN 1-880618-01-X : $24.95. -- ISBN 1-880618-00-1 (pbk.) : $16.95
   1. Woodwork. 2. Wooden toy making. 3. Furniture making.
I. Woodworker's journal.
TT180.W647   1992
684'.08--dc20   91-38734
                                                              CIP

Madrigal Publishing Company
517 Litchfield Road
P.O. Box 1629
New Milford, CT 06776

Contents

Introduction 1
Mitered-Corner Box 2
Puss 'n Books 6
Pierced Tin Cabinet 7
Cookbook Holder 12
Wooden Jewelry 14
Child's Duck Puzzle 16
Country Occasional Table 18
Early American Candlesticks 20
Arrow Wall Decoration 22
Three-Drawer Country Wall Box 24
Key Cabinet 26
Contemporary Box 28
Glass-Top Table 31
Shaker Carrier 34
Hunt Table 36
Knife and Fork Tray 39
Loon Carving 42
Old-Time Pipe Box 48
Dutch Tulip Folk Art Silhouette 50
Four-Drawer Lamp 51
Salt Box 54
Bud Vase 56
Early American Dry Sink 58

(continued on next page)

Oak Magazine Rack 62

Key Spline Jig 64

Shaker End Table 65

Heart Stool 68

Decorative Cutting Boards 70

Piggy Bank 72

Turned Bowl 76

Country Cupboard 78

Coffee Table 82

Rooster Folk-Art Silhouette 86

Harvest Basket 88

Bird Push Toy 90

Pencil Post Nightstand 93

Vegetable Bin 96

Treetop Christmas Ornament 100

Classic Pickup Truck 102

Whale Pull Toy 106

Child's Carousel Lamp 109

Colonial Wall Sconce 112

Sources of Supply 113

Acknowledgments

Our thanks to the following individuals who provided material for this book:

Skip Arthur, Saugerties, N.Y., for the Whale Pull Toy and the Bird Push Toy; **Brian Braskie**, Canterbury, N.H., for the Shaker End Table; **Rick and Ellen Bütz**, Blue Mountain Lake, N.Y., for the Loon Carving; **Fred Cairns and Kathy Dawson**, Greenleaf, Kans., for the Classic Pickup Truck; **Gene Cosloy**, Wayland, Mass., for the Country Coffee Table; **Robert Leung**, Benicia, Calif., for the Contemporary Box; **Tony Lydgate**, Palo Alto, Calif., for the Mitered Corner Box; **Donald McLean**, Manhasset, N.Y., for the Cookbook Holder; and **Joe Robson**, Trumansburg, N.Y., for the Shaker Carrier.

Also, many thanks to **John Kane** of Silver Sun Studios, New Milford, Conn., for the cover photo and many of the project photos.

The Editors

Introduction

Worldwide, over the past decade, woodworking has seen tremendous growth. In this electronic age, where so much of what we do is one-dimensional, there seems an ever-stronger urge to engage in hobbies where the result is three-dimensional. The need to create something real, something that we can feel and hold, is undeniable. It affirms our creative urges, soothes our souls, and helps to validate our existence.

But as anyone who works with wood quickly discovers, the keys to success in woodworking are good designs and accurate plans. For 15 years, as the editors of *The Woodworker's Journal* magazine, we've been providing woodworkers with the plans they need. Now, in *The Woodworker's Project Book*, we've compiled 42 of our best plans. Each project was selected with an eye toward three criteria: It had to be a good design, it had to be well-constructed, and it had to be easy-to-make.

The Woodworker's Project Book has something for everyone. There are easy weekend projects, such as the Cookbook Holder, the Country Wall Box, and the Heart Stool. There are just-for-fun projects like the Kid's Piggy Bank, the Harvest Basket and the Rooster Folk-Art Silhouette, and there are toy projects like the Classic Pickup Truck and the Whale Pull Toy. But *The Woodworker's Project Book* is much, much more. It's some of the best in period and country furniture, with the Shaker End Table, the Pencil Post Nightstand and the handsome Early American Dry Sink. And it's some of the best in gift projects, with the Child's Carousel Lamp, the Mitered-Corner Box, and the elegant Wooden Jewelry. There are even turning projects, like the Bud Vase, and the Turned Bowl, and carving projects, like the beautiful Loon.

Sometimes a book can be more than the sum of its parts. We think you'll find that *The Woodworker's Project Book* is more than just a collection of 42 fine projects. It's an opportunity to share a rewarding hobby, where the fruits of your labors can be seen and touched. And it's an opportunity to share a very special feeling—the feeling of pride when you can say "I made that."

The Editors

MITERED-CORNER BOX

by Tony Lydgate

This piece was designed to meet the need for a box that's fairly simple and inexpensive to make. The challenge was to generate a design that was interesting, and that would best take advantage of the coloring and contrast possibilities of a variety of exotic hardwoods. I usually craft these boxes from cocobolo, bocote, curly koa, and figured maple.

I opted for a basic square form, 4⅜ in. on a side and 2¾ in. high. Start by getting out stock for the sides (A). Since it's just as easy to make four boxes as it is to make one, I recommend that you start with a ¹³⁄₁₆ in. thick by 2¼ in. wide by 38 in. long board which, when resawed and crosscut, will yield enough stock for the sides of four boxes. If you're not experienced working with small pieces on the table saw, you'll feel more comfortable using this longer board for the initial resawing operation. Depending on your saw's capacity, you may need to use several blade height settings to resaw this 2¼ in. wide stock. Resawing on the band saw, though less precise, is one other option you might want to consider.

Before resawing, I sand the two faces of the board on my 6 by 48 in. belt sander. I start with a 60-grit sandpaper then move up to 120-grit, and finish with what I call a polishing belt, which is a 120-grit belt that's been used so long it has no "teeth" left, but imparts a nearly glass-smooth polish to the stock. Since the outside faces of this board, after milling, will become the inside faces of the box sides, this step eliminates the need for further sanding of the interior.

I now set up my table saw (with a top quality rip blade) to resaw the board to ⁵⁄₁₆ in. thickness (Step 1). I set the fence so I'll get slightly (¹⁄₃₂ in.) over ⁵⁄₁₆ in. to avoid any possibility of the side walls being too thin. Use a push stick and keep your hands clear of the blade. Then I cut the ³⁄₁₆ in. deep groove into the two resawed blanks that will accept the ⅛ in. thick plywood bottoms (B), making certain that these grooves are milled into the sanded or inside face (Step 2). Next crosscut the two resawed blanks into 4½ in. long pieces, slightly more than the 4⅜ in. finish length of the sides (Step 3).

Now set your miter gauge at 90 degrees, tilt the crosscutting blade to 45 degrees, and position a stopblock on the miter gauge fence so that you'll end up with sides that are exactly 4⅜ in. long (Step 4). It's best to take some scrap stock and cut four test pieces to check your setup before going to work on the project stock. If the miters of your test box are

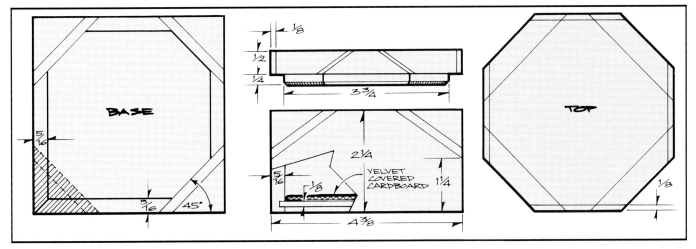

off, then either the miter gauge is not at exactly 90 degrees, or the blade setting is not at precisely 45 degrees. Readjust them as needed. After a few minutes work, you'll have a stack of sixteen perfectly uniform side parts.

For the bottoms (B) I use lauan mahogany, which costs about $10 for a 4 by 8 sheet. Of course, you can use any plywood that you have around the shop; just be sure that you machine the grooves in the sides equal to the thickness of the plywood you select. Make the plywood bottoms a hair smaller than the actual groove-to-groove dimensions to allow a fudge factor during glue-up. When assembling, keep the good side of the plywood down; the bad side, which is the inside of the box, will be covered by a velvet liner.

You'll need some 2 in. wide masking tape and a square for gluing up the boxes. The masking tape will serve as your clamps. For small assemblies such as this box, where traditional clamps would be cumbersome, the masking tape allows for quick and easy assembly with a minimum of fussing. Start by laying out the four sides, with two of the sides each positioned on a section of masking tape about 7 in. long, as shown in Step 5. Now, using a glue bottle with a narrow spout, run a bead of glue into the bottom groove in all four sides, and onto the miters on the two sides without tape. How much glue should you use? Well, too little and the box won't hold together; too much and the excess glue will squeeze out. If some glue squeezes out, either wipe the excess off immediately with a damp cloth, or wait 30 minutes until the glue is chewy, and then chisel it off.

To assemble the boxes, first insert the bottom into one of the sides without tape, then add a side with tape, butting the mitered joint tightly together with finger pressure, and wrapping the tape around to securely clamp the joint. Next, add the third side (without tape), and then the last side (with tape), wrapping the tape tightly around the joints to secure them (Step 6). Now place the box on a flat and true surface, such as the saw table, and use the square to insure that all corners are 90 degrees.

While you are waiting for the boxes to dry, you can start making the tops, which are ½ in. thick hardwood (C) with a ⅛ in. applied edging (D) in a contrasting wood. Since you are making four boxes, you'll also need to make four tops. Again, it's always safer to work with longer pieces of wood. I cut the edging first, ripping 20 in. lengths of ⅛ in. edging from 8/4 stock, and then ripping to width. For these ripping cuts, you'll need to use a plywood table saw insert, with the ripping blade raised up to cut through the insert. This will eliminate the gaps on either side of the blade that with conventional metal inserts pose a hazard of swallowing narrow ⅛ in. edging.

The hardwood board from which the lids will be crosscut must be ½ in. thick by 4⅛ in. wide by about 20 in. long. Glue the 20 in. strips of edging along either edge of this board, using masking tape every few inches instead of clamps (Step 7). When the glue has dried, remove the masking tape, and sand both faces of the edged board with the same sanding process you followed on the box sides. Then crosscut into four pieces, each piece measuring 4⅛ in. long. Once crosscut, add the ⅛ in. edgings on each end (Step 8). Again, use glue and masking tape instead of clamps to apply this edging.

The mitered corners or "dog-ears" of this box are its most appealing design feature. They are cut in a two-step process, with a roughing cut made on the band saw, and the finish cut done on the table saw. The band saw cut is made with the saw

(continued on next page)

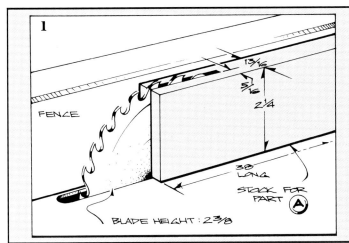
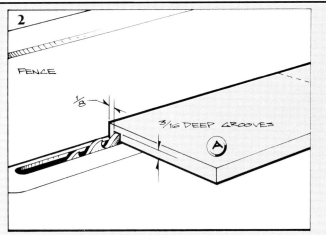

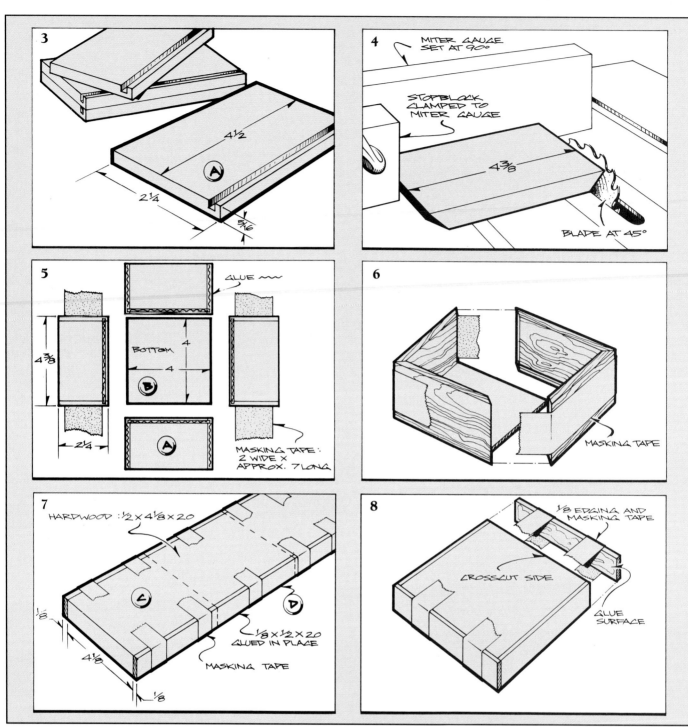

table tilted at 45 degrees, and the miter gauge also set at 45 degrees. As shown in Step 9, you'll need to add a stopblock to the miter gauge fence and turn the boxes upside down for this operation. The roughing cut should remove the corner starting at a point just about 1⅜ in. up from the bottom edge of the box. After cutting all four corners on the band saw, move to the table saw. The finishing table saw cut, made with the blade angled at 45 degrees, and the miter gauge angled at 45 degrees and equipped with a stopblock (see Step 10), is basically just a clean-up cut. Don't try to get the setup right with the first cut. It's best to start with the stopblock a little farther away, then gradually move it in until the cut made starts exactly 1¼ in. up from the box bottom.

Once you have trimmed the four corners on the table saw, you should have a perfectly smooth and flat surface on which to glue the triangular dog-ears (E). Rip some contrasting stock about ⁵⁄₃₂ in. thick and 1¾ in. wide, then cut the stock on the band saw into triangles large enough to fit over the corners with at least ⅛ in. overlap all around. I run the grain horizontally, though you could make it vertical if you prefer. I apply aliphatic resin (yellow) glue to the ears, and then hold them in place with steady finger pressure till the glue has started to set (Step 11). No clamps are needed. When the glue has dried, I simply sand the overhang flush on the belt sander.

While you are waiting for the ears to dry, go back to work on the lid. Use the same band saw setup that you used in cutting the box corners to cut the corners on the lid. Set the stopblock so that your cut will be just shy of the final desired mark that would carry the plane of the box corner through to the lid.

Next, rip ¼ in. thick stock for the lid liner (F). You'll need to rip, crosscut, and then nip the corners using the table saw miter gauge until the lid liner fits snugly inside the box.

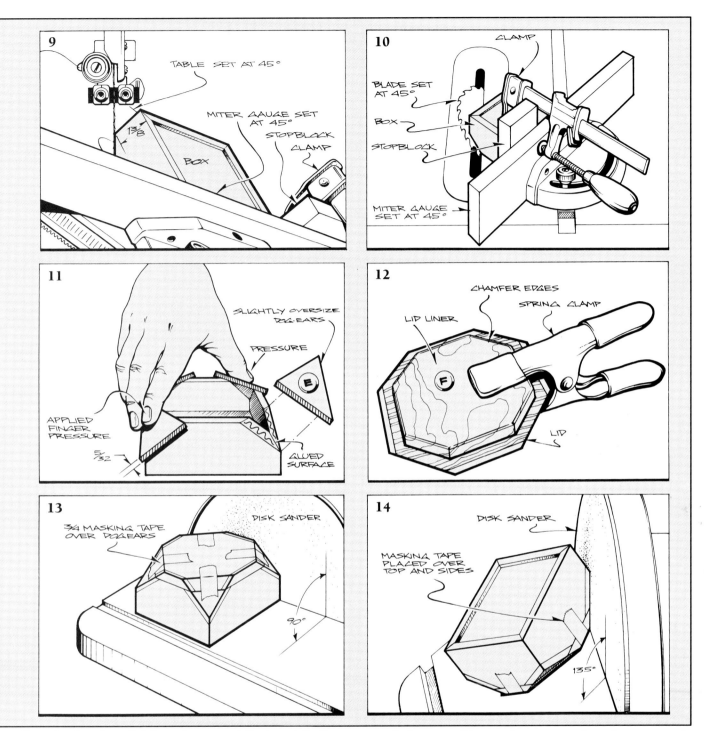

Chamfer the edges slightly as shown, and final sand the liner on what will be its exposed face.

To fit the liner to the top of the lid, spread glue on the underside of the liner, and position it on the top using your eye as a guide. Move the top about a little, working the glue in until the two pieces start to tack. Now final center the liner on the top by test-fitting the lid. Wait a few moments, then very carefully remove the lid and use spring clamps to clamp the liner to the top (Step 12). Make sure the liner does not shift position under the pressure of the spring clamps.

To final shape the box, first tape the lid to the box with ¾ in. masking tape applied across the ears, as shown in Step 13, so the four flat faces of the box are free of tape. Sand the sides of the boxes on your disk sander and then your belt sander, using the same sanding steps that you followed earlier. Now apply masking tape on the just sanded sides, and then remove the tape from the ears. This way the lid won't change position, and you can perform the same sanding operation on the ear surfaces (Step 14). Don't be discouraged if you don't get the angles on the first box exactly right. That's another strong argument for making four boxes; by the fourth box you should have the technique down perfectly.

To finish, oil the box with Watco or a comparable oil finish, wet-sanding the box at the same time with steel wool or a wet-or-dry paper. When dry, rub with 000 steel wool to remove the excess finish, and then wax.

Cut a piece of mat board to fit inside the box, and wrap it with your choice of velvet, using masking tape to hold the velvet on the underside of the mat. Some velvet is treated with silicone and tape won't stick to it, so you should bring tape with you to the store and check the product before you buy. Drop the velvet liner in after the waxing is done. I generally sign these boxes on the underside of the lid.

Our cat bookends are the perfect afternoon project; simple to make and a sure attention grabber. To make one set, you'll need a ¾ in. by 5½ in. by 30 in. long pine or hardwood board, 4 screws, and white, black, and green enamel.

Use the full-size pattern with a sheet of carbon paper to lay out the cat profile on one end of the board. Also lay out the two backs (A). With the table or radial-arm saw, crosscut the cat profile at the cut line (see full-size pattern), then use a jigsaw or saber saw to cut all the parts out.

Final sand the cat to remove saw marks, then paint it using the color scheme indicated on the full-size pattern. Before finishing the backs with lacquer, we drilled and countersunk for the flathead brass wood screws that are used to join the back parts to the front and rear halves of the cat. Since our bookends were intended as a light-duty decorative piece, we did not add a plate under them.

You should add a 4 × 5 in. steel plate under each bookend if you plan to use them to hold a large number of books.

Puss'n Books

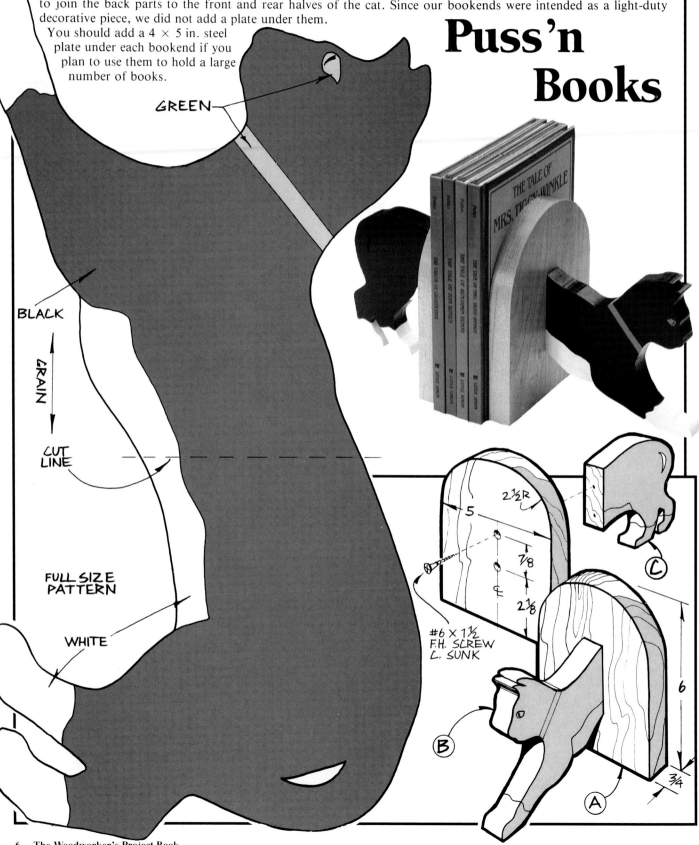

Pierced Tin Cabinet

This handsome Pierced Tin Cabinet is based on similar cabinets that were popular in the 18th century. To simplify making the cabinet we've included step-by-step instructions for piercing the tin, and we've listed a mail order source where you can purchase either a blank or pre-pierced tin panel and tools for piercing.

To minimize the chance of wood movement problems we used ¼ in. thick birch plywood for the back (J). If you don't plan to paint the cabinet, though, knotty pine plywood would be a better choice.

All other parts are made from solid pine. Try to select stock with a minimum of knots as they reduce strength. Also, unless the knots are treated with a wash coat of shellac, their resin may bleed into a painted surface.

The two sides (A) can be made first. Cut ¾ in. thick stock to 6½ in. wide by 22¾ in. long, taking care to insure that each end is cut square. The ¾ in. wide by ¼ in. deep dado that will accept the bottom (C), and the ¾ in. wide by ¼ in. deep rabbet that will accept the top (D), can be cut using the dado-head cutter in conjunction with the table saw and a miter gauge. Once the rabbets and dadoes are cut, lay out and mark the location of the ¼ in. diameter by ½ in. deep dowel holes, then use a drill press to bore them out.

To make the base (B), cut ¾ in. thick stock to 7½ in. wide by 16¾ in. long, again making sure that the ends are square. Next, lay out and mark the location of the ¾ in. wide by ¼ in. deep stopped dado as shown. To cut the dado you'll need a router equipped with an edge-guide and a ¾ in. diameter straight bit. For a smooth cut — and to minimize strain on the router — it's best to cut the ¼ in. depth in two passes, each pass removing ⅛ in. of stock. If your router bit collection does not include a ¾ in. straight bit, readjust the guidefence to get the ¾ in. dado width. Keep in mind that the router bit will leave rounded corners at the stopped ends, so once the dadoes have been cut, you'll need to use a chisel to square the corners.

Note that the base has a bead cut on each end and on the front edge. This can be cut using a router and a ½ in. bearing-guided beading bit. Set the bit to a depth that will create the ⅛ in. step, then cut the bead on each end of the base. Once the ends have been routed, cut the bead along the front edge. Routing the front edge last will clean up any splintering that may have occurred when the ends were cut.

Next, from ¾ in. thick stock, cut the bottom (C) to 6¼ in. wide by 13¾ in. long and cut the top (D) to 6½ in. wide by 13¾ in. long. As with the other parts, check the ends for squareness.

Now final sand the sides, base, bottom and top. Start with 80-grit sandpaper, then follow with 100-, 150-, and 220-grit to complete the sanding. When sanding the ends of the bottom, keep in mind that if too much material is removed you'll no longer have a snug fit in the side dadoes. Try to maintain a snug fit to get maximum strength from the joint.

The sides, base, bottom, and top can now be assembled. Apply glue to all mating surfaces, then clamp firmly and check for squareness. Allow to dry thoroughly.

When dry, drill counterbored holes for 1½ in. by no. 8 flathead wood screws as shown. Drive the screws, then plug the holes with short lengths of dowel stock. Cut the dowels a bit long so that after they are glued in place they can be sanded flush with the surface.

The ⅜ in. wide by ¼ in. deep rabbet for the back can now be cut using a router equipped with a ⅜ in. bearing-guided rabbeting bit. Set the bit to make a ¼ in. deep cut, then with the base of the router against the back edge of the case, rout the rabbet all around as shown. Since the bit will leave rounded corners, you'll want to use a chisel to square them.

Next, cut the ¼ in. thick plywood back to fit, then attach it to the case with glue and ½ in. by no. 4 flathead wood screws. Since the cabinet will be hung by driving screws through the back, be sure the back is well secured to the cabinet.

(continued on next page)

Bill of Materials
(all dimensions actual)

Part	Description	Size	No. Req'd.
A	Side	¾ × 6½ × 22¾	2
B	Base	¾ × 7½ × 16¾	1
C	Bottom	¾ × 6¼ × 13¾	1
D	Top	¾ × 6½ × 13¾	1
E	Filler	¾ × 2 × 13¼	1
F	Adjustable Shelf	¾ × 5½ × 13⅛	1
G	Side Molding	see detail	as req'd.
H	Front Molding	see detail	as req'd.
I	Cleat	¼ × 1⅛ × 6½	2
J	Back	¼ × 14 × 22½	1
K	Turnbutton	¼ × ¾ × 1	1
L	Door Rail*	¾ × 1½ × 13¼	2
M	Door Stile	¾ × 1½ × 16¼	2
N	Tin	28 Ga. × 11 × 14	1
O	Keeper Strip	¼ round	as req'd.
P	Drawer Front	¾ × 4 × 13¼	1
Q	Drawer Side	½ × 4 × 6	2
R	Drawer Back	½ × 3½ × 12¾	1
S	Drawer Bottom	¼ × 5¾ × 12¾	1
T	Drawer/Door Knob	¾ dia. (porcelain)	2
U	Hinges	1½ × 1½	1 pair

*Includes tenons.

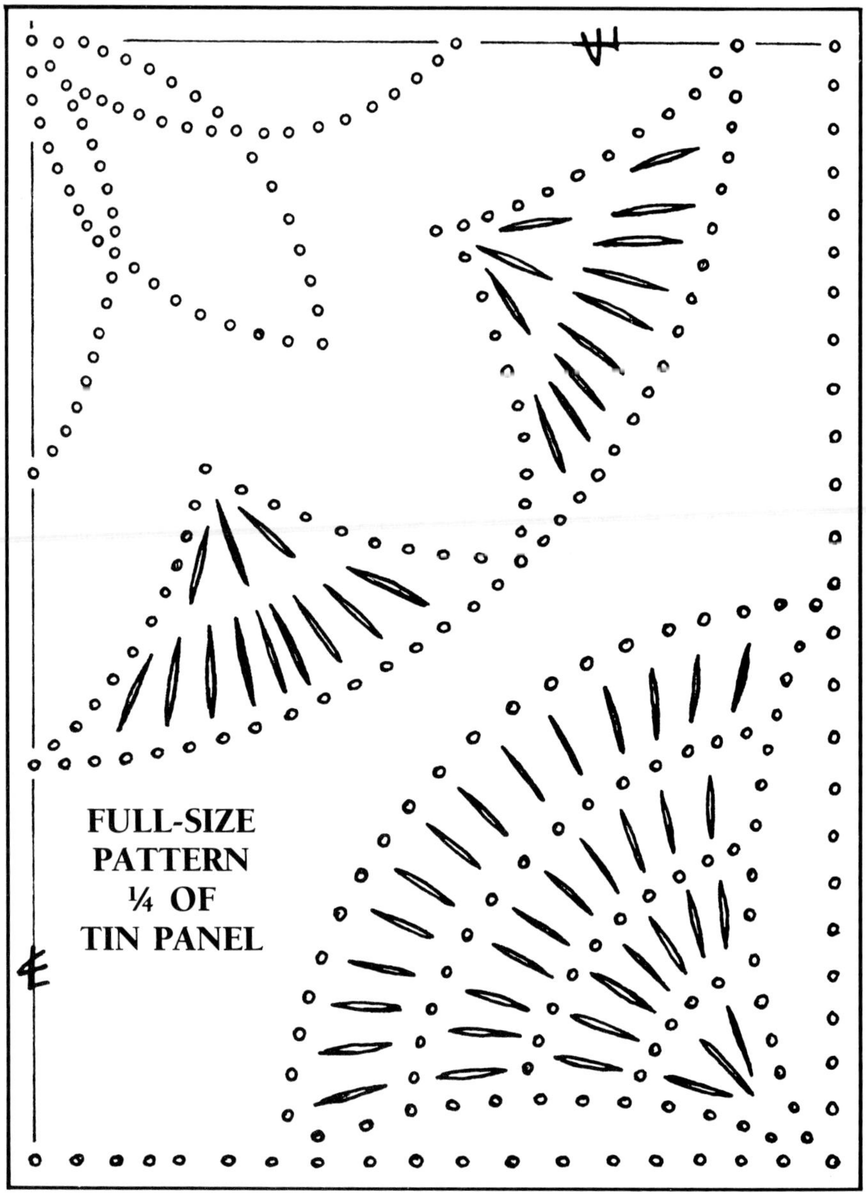

FULL-SIZE PATTERN ¼ OF TIN PANEL

Make the door next. Cut the rails (L) and stiles (M) to overall length and width from ¾ in. thick stock. We used a tenon jig to cut the tenons on each end of the rails and the through-mortises on each end of the stiles. Once the joints are cut, apply glue to the mating surfaces and clamp firmly. Check for squareness before setting aside to dry. When dry, use the router and a ⅜ in. bearing-guided bit to rout the ⅜ in. by ⅜ in. rabbet all around the inside edge as shown. Once cut, use a chisel to square the rounded corners.

Prepare the tin panel (N) as described in the Pierced Tin Step-by-Step instructions on page 10. It's held in place with 1/4 in. quarter-round molding (O) tacked in place with small brads. The molding may tend to split when the brad is driven. To prevent this, it's a good idea to first bore a small pilot hole for each brad.

The drawer is made as shown. Note that the drawer front (P) has a ½ in. by ½ in. rabbet while the drawer sides (Q) have ¼ in. by ½ in. dadoes. The bottom (S) can be made from the same ¼ in. thick plywood used for the cabinet back.

You can get the crown and cove molding (see sectional detail) for parts G and H at most lumberyards and building supply centers. To eliminate any movement problems between parts A and parts G (parts A will expand and contract, front to back, with changes in humidity while parts G will not), we've worked out a method of joinery that allows the inevitable movement, yet securely attaches the side moldings. The groove for the cleat (I) is cut on the table saw (see detail), then the cleat is cut to size and glued in place. At the front, about 1 in. from the miter, bore a 5/32 in. diameter hole (not shown on drawing). Following this, cut a pair of 5/32 in. wide by ⅜ in. long slotted holes, one in the middle and one about ¾ in. from the back end (see exploded view).

To attach each side molding, drive ¾ in. by no. 6 flathead woodscrews through the three cleat holes and into the top edge of the side. Do not use any glue here as the side must be free to expand and contract.

Since the grain of the front molding (H) runs in the same direction as the top, there is no need to add a cleat. It can simply be glued in place and secured with countersunk and filled finishing nails.

Next, the filler (E) can be cut to size and glued in place. Use clamps to secure it while the glue dries. When clamping, make sure that the front edge of the filler is flush with the front edge of the top.

The cabinet will look good simply stained and finished with a varnish or penetrating oil. However, since a good many Early American pieces were painted, we felt that a milk paint finish would add authenticity. Milk paint can be ordered from The Old Fashioned Milk Paint Company, Box 222, Groton, MA 01450 (telephone: (617) 448-6336). We used their "Soldier Blue," although several other antique colors are also available. The addition of the porcelain knobs (T) and the hinges (U) will complete the project.

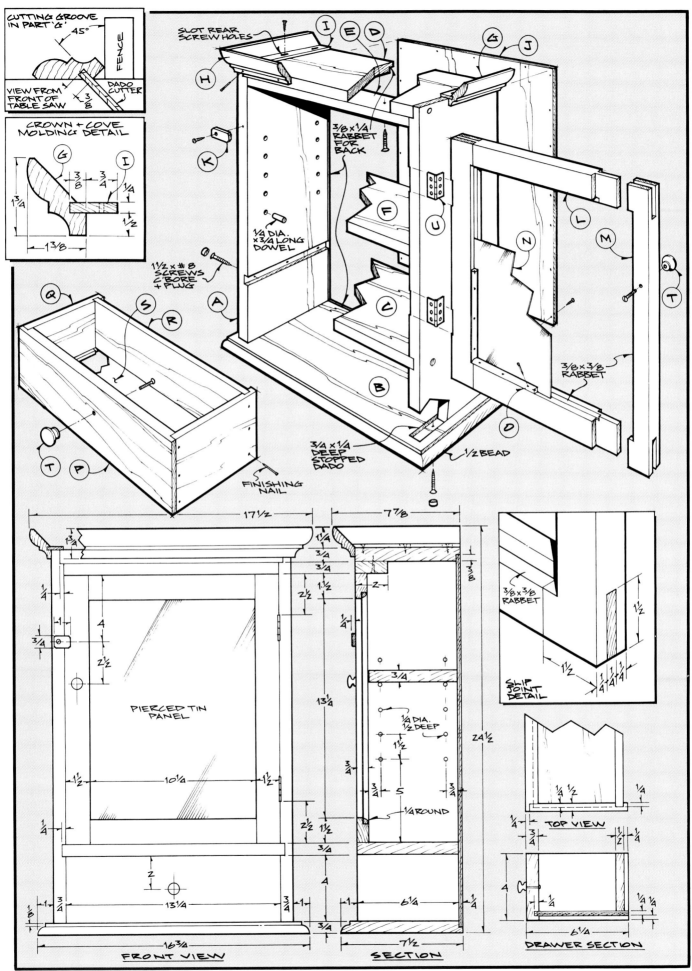

Pierced Tin: Step-by-Step Instructions

Tin piercing is a centuries-old art form, practiced in the Middle Ages, and brought to the colonies by the early settlers. Pierced tin foot warmers, lanterns, and other decorated tinware were common household items in Colonial and Early American homes.

The pierced tin panels most often associated with woodworking were used primarily in cabinets and pie safes. The perforations that created patterns on the tin panels served the dual purpose of allowing air to circulate to baked goods, while keeping flies and mice out.

There are many types of tin sold today for the purposes of decorative piercing. Among the most popular are antiqued tin, pitted and rusty tin, old-look tin, pewtertone tin, and bright tin. We used bright tin because it was the least costly and most attractive for our project.

In addition to the tin panel, you will need tools for piercing. (See editor's note at the end of this article.) We used only two punches, an all-purpose point punch and a ½ in. lampmaker's chisel. While a nail set and an old chisel could be used instead, the professional punches result in a higher quality, more consistent look, and are therefore recommended.

Rounding out your materials for punching will be a firm work surface, an 8-or 12-ounce ball peen hammer, a 2 ft. by 2 ft. by ¾ in. thick sheet of smooth plywood or particleboard, a rawhide mallet, a straightedge, some paper, pushpins, tape, and a can of spray lacquer or polyurethane. It's usually best to transfer the pattern directly to the tin, and if you were doing so, you would need a felt-tip pen or grease pencil and a compass. However, with the full-size one-quarter pattern provided, we've worked out a method where no drawing is necessary. Our method involves taping the pattern to the tin, although there is always some danger of the paper slipping out of alignment, which could result in an uneven pierced tin pattern. We urge you to check the paper alignment periodically while punching to insure that no movement has occurred.

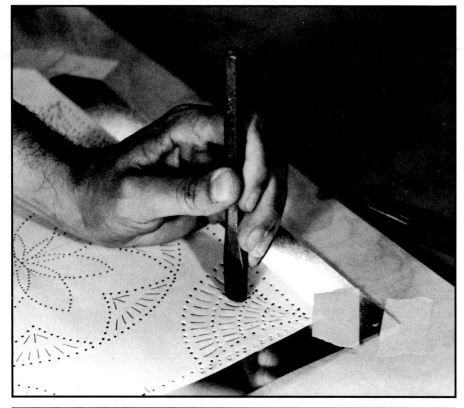

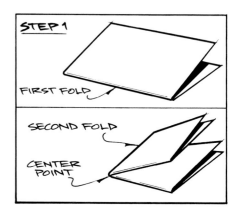

Step 1: In order to make a full pattern from the one-quarter section provided, first take a piece of paper about 11 in. by 14 in. and fold it in half, top to bottom, and in half again, side to side. Mark the inside corner (this will be the pattern center point when the paper is unfolded). Make a copy of our one-quarter pattern on a photocopier, or by using carbon paper and a blank sheet of paper. You could simply use the pattern on the page, but since you will be piercing through, you would end up with a perforated page.

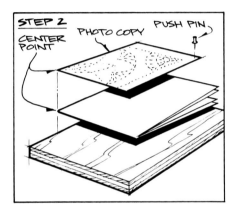

Step 2: Trim the copy of the one-quarter pattern along the center lines as shown, then position the folded paper under the pattern with the ¾ in. thick plywood below. Note the center point of the paper in relation to the pattern. Anchor the pattern and paper to the plywood with push pins as shown.

Step 3: Pierce through the paper. The point punch is used for the round holes, and the lampmaker's chisel for the elongated perforations. Remove the pushpins and unfold the paper.

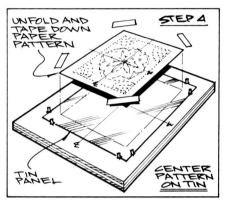

Step 4: Position the pierced paper pattern so that it is centered over the tin panel, and tape it in place. Use pushpins at the four corners to anchor the tin panel to the plywood backing. Take extra care when handling the tin, both from a safety standpoint and to avoid smudging it. A pair of light cotton gloves will prevent finger smudges (which "etch" some tin) and cuts from the tin's sharp edges. You can also wrap the edges of the tin with masking tape to prevent accidental cuts while handling.

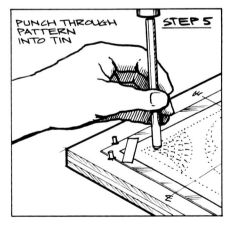

Step 5: Pierce through the pattern paper and tin, using the same punches as before. If you have a scrap of tin or an extra practice piece, try working out a small section of the pattern first to gain experience before committing your project stock. It is best to get a feel for using the punches since the velocity and weight of the hammer can have a varying effect and result in irregularly sized holes which tend to look rather sloppy.

Step 6: Remove the tacks, flip the pierced tin panel over, and use the rawhide mallet to gently flatten out the dents. Seal the tin with a spray lacquer or polyurethane. This is important since it will preserve the appearance of the tin, preventing it from rusting or oxidizing, and it will make for easy cleanup. A satin, matte, or gloss finish will do, depending on personal taste. Three or four light applications are always better than one or two heavy coats. Remember to use an even sweeping motion when spraying, and to let each coat dry thoroughly before applying the succeeding coat.

Pierced tin panels that have been sprayed with a protective lacquer or polyurethane are simple to keep clean, requiring only a minimally damp cloth to wipe down occasionally. Do not use abrasive cleansers or rub the panels with oil. While oil was once used to prevent oxidation, it is a magnet for dust and unnecessary with modern lacquer and polyurethane finishes.

Old-look tin can be purchased ready to use, or you can dull the look of the bright tin yourself by rubbing out the surface with a fine Scotch pad or 0000 steel wool, and brushing on a coat of naval jelly. The longer the naval jelly remains on the tin before it is washed off, the duller the resulting look.

Editor's Note: Our thanks to Jim and Marie Palotas of Country Accents for their help with this article and for the use of the pattern presented here. Country Accents offers a complete line of tin, copper and brass panels, a wide variety of punching tools, and hundreds of different patterns. The 11 in. by 14 in. bright tin blank panel required for our cabinet, or a pre-pierced panel, can be purchased from them. The point punch (part no. T-0359), the ½ in. lampmaker's chisel (part no. T-0259), the rawhide mallet and the ball peen hammer are also available. For more information or to order, write to: Country Accents, P.O. Box 437, Montoursville, PA 17754. Telephone: (717) 478-4127.

Adjustable Cookbook Holder
by Donald McLean

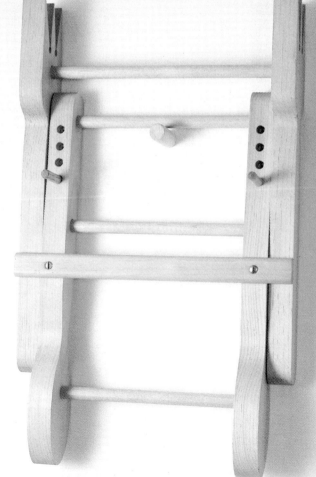

This handy kitchen item keeps a cookbook open to the proper page while holding it at an angle that makes for easier countertop reading. And when you are finished, it folds flat to fit in a drawer or, if you prefer, it can be hung nearby where it will double as a wall decoration. The one shown is made from pine, but maple, ash, or oak also come to mind as good choices.

To make the forks (A) and spoons (B) you'll need four pieces of 3/4 in. thick stock, each measuring 2 in. wide by 12 in. long. Referring to the grid pattern on the drawing, transfer the profiles to the stock. At the same time, mark the location of the 3/16 in. diameter holes in the fork tines. Bore the holes, then use the band saw to cut out the profiles. When making the band saw cuts, stay about 1/16 in. on the waste side of the stock. After the cuts are completed, use a file and sandpaper to smooth the edges exactly to the marked lines.

Next, lay out and bore the 3/8 in. diameter holes for the two fork stretchers (D) and the two spoon stretchers (E). Also bore the 1/4 in. diameter holes for the pegs (F).

After cutting the stop (C) to size, the project is ready for final assembly. First, though, give all parts a thorough sanding, finishing with 220-grit paper. Apply glue to the ends of the two spoon stretchers, then add the spoons and clamp firmly. When dry, glue the stop in place using 1/2 in. long by no. 6 oval-head wood screws to reinforce the joint.

Glue one end of each fork stretcher into one of the forks. Allow to dry, then assemble the spoon frame (parts B, C, and E) and add the second fork. When dry, final sand all parts and finish with two coats of penetrating oil.

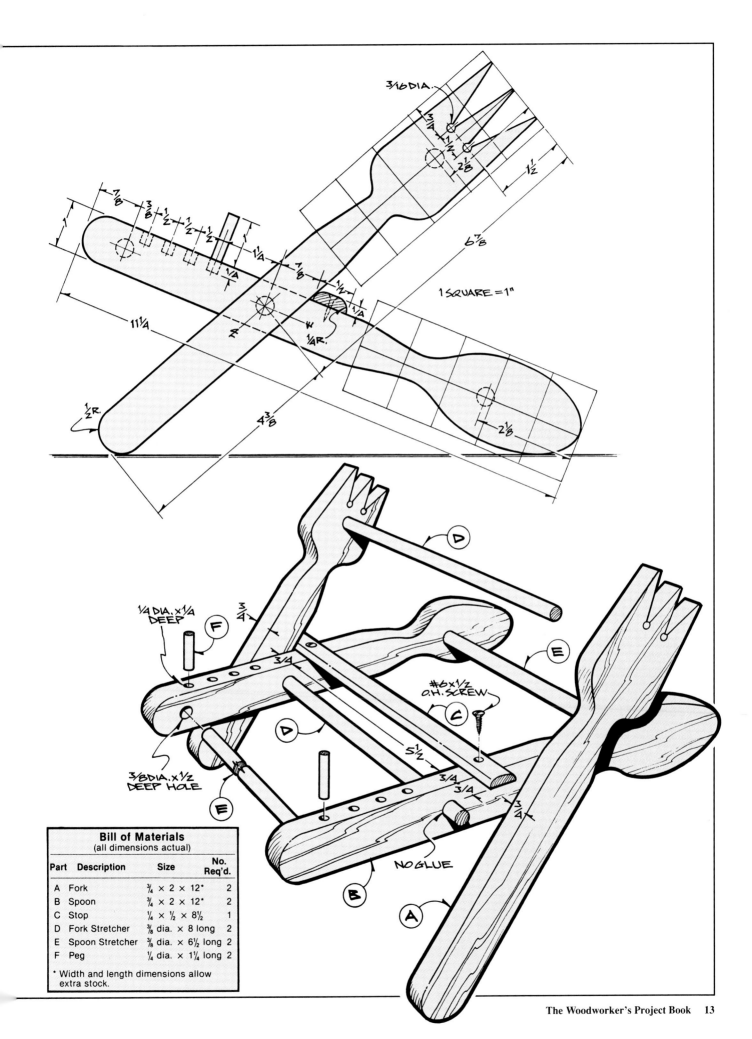

Wooden Jewelry

With a holiday or anniversary usually just around the corner, this jewelry might make the perfect gift for someone special.

We chose ebony, satinwood and bloodwood for our jewelry, but you may want to experiment with other woods. A broad variety of hardwoods, including those we used, are available from: The Berea Hardwoods Co., 125 Jacqueline Drive, Berea, OH 44017.

To start, cut the five wedges that comprise the block from which the pin and earrings are resawed. Make the cutting fixture shown in Step 1 to hold the 1 in. thick blocks of each wood. We used five wedges at 18 degrees each to make up the 90 degrees; you could use any other number of wedges, so long as the total adds up to 90 degrees and all the wedges are equal. The notch in the cutting fixture must be cut equal to the wedge angle — 18 degrees in our case.

To make the jewelry exactly as we have, you will need one block each of ebony, satinwood and bloodwood, cut to the dimensions shown in Step 2. After cutting the first wedge, reverse the stock in the jig to get the second wedge out of the satinwood and bloodwood blocks. As shown in Step 3, the cleat anchors the stock securely. Set the fence so the blade just kisses the edge of the fixture. Raise the blade to a 1⅝ in. height, and keep your hand well away from the blade while cutting.

To glue up your five 18-degree wedges, you'll need to make a simple fixture to hold them in the proper alignment. As noted in Step 4, the angle of the fixture should be exactly equal to the total of your wedges. Position the wedges on the fixture block and mark the outside edges to get the proper angle. Although a 90-degree angle should work, the sum of the wedges will probably be a little off, since it is difficult to achieve a perfect 18-degree cut.

Arrange the wedges as shown in Step 8 with respect to the order of the different types of wood, apply glue, and use a band or belt clamp to apply pressure. A C-clamp, with waxed clamp blocks top and bottom, will keep the wedges flush as the belt clamp is tightened (Step 5).

When dry, resaw the glued-up block into three ¼ in. thick sections, using the band saw as shown in Step 6. Hand sand (Step 7), and then lay out the pin and earring profiles on the ¼ in. thick sections. The earrings are laid out on one section, the pin on another, and the alternate pin profile on the third. You could also use the third block to make an extra set of earrings or some other article of jewelry.

Next, cut out the profiles (Step 9) with a scroll saw, or by hand with a coping or jeweler's saw. To resaw the earrings, use a dovetail saw, first cutting halfway through (Step 11A), then inserting a slip of paper in the kerf to complete the cut (Step 11B). Use a half-round and a flat file to smooth the edges before final sanding. Mount the earring posts and pinback with epoxy glue (Step 12) before finishing with four light coats of a gloss or semi-gloss aerosol spray lacquer. The 1¼ in. long pinbacks and surgical stainless steel 5mm flat pad earring posts are available at most craft supply stores, or they may be ordered from the National Artcraft Co., 23456 Mercantile Road, Beachwood, OH 44122.

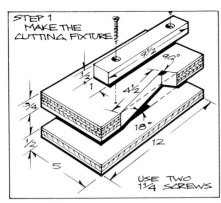
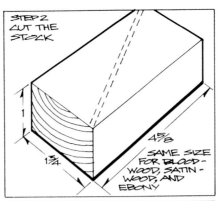
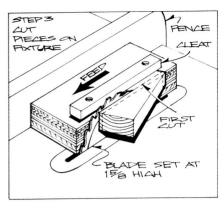
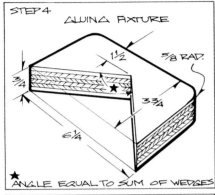
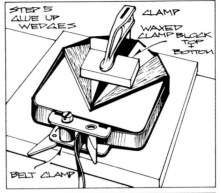
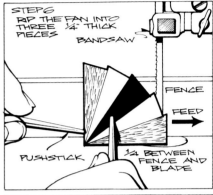

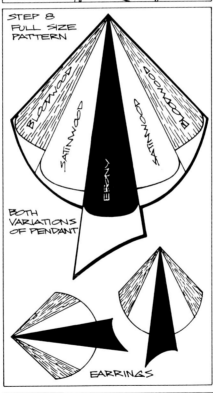
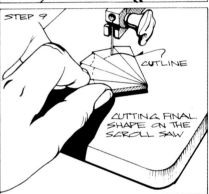
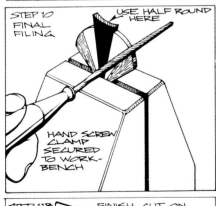
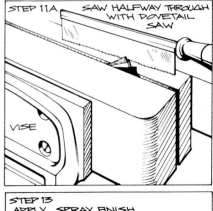
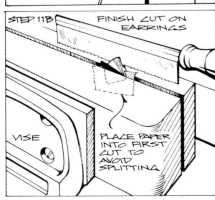
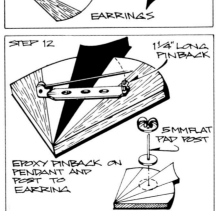

The Woodworker's Project Book 15

Child's Duck Puzzle

Children always seem to enjoy the challenge of a puzzle, and we think this one, designed for two- to four-year olds, should be no exception. The base and the puzzle parts are made from ¼ in. thick birch plywood, while the edging is ½ in. thick solid maple. Since small parts present a choking hazard to young children, we designed the puzzle making sure that all parts are big enough to be safe.

Start by cutting the plywood to 7¼ wide by 10 in. long, then transfer the full-size duck pattern shown. Note that the double line represents the cutting line for each of the puzzle parts. Use either a band saw equipped with a ⅛ in. blade or a scroll saw to make the cuts.

Now paint the puzzle parts. Keep in mind, though, that young children sometimes chew on toys, so it's important to use paints that are non-toxic. If not available locally, the company Cherry Tree Toys, P.O. Box 369, Belmont, OH 43718 sells non-toxic, semi-gloss enamel paint in a variety of colors. You'll need five: red, blue, green, yellow and white. Orange is a mixture of three parts yellow and one part red, light blue has three parts white and one part blue, and light green consists of three parts yellow and one part green.

Next, cut the plywood base and edging, then assemble as shown. Final sand, taking care to remove any sharp edges. No final finish is needed.

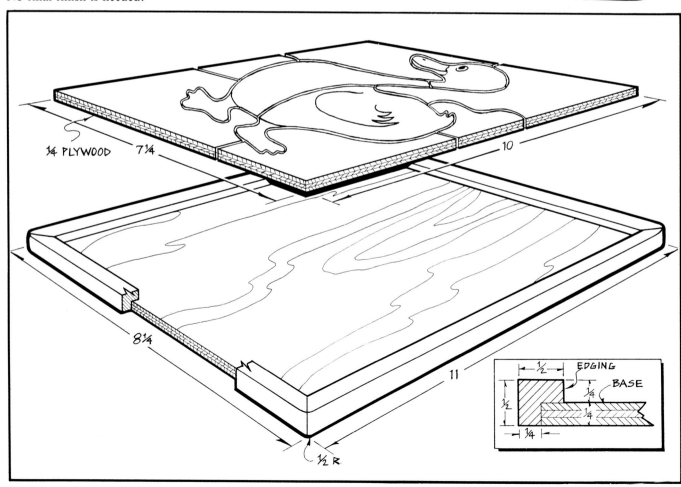

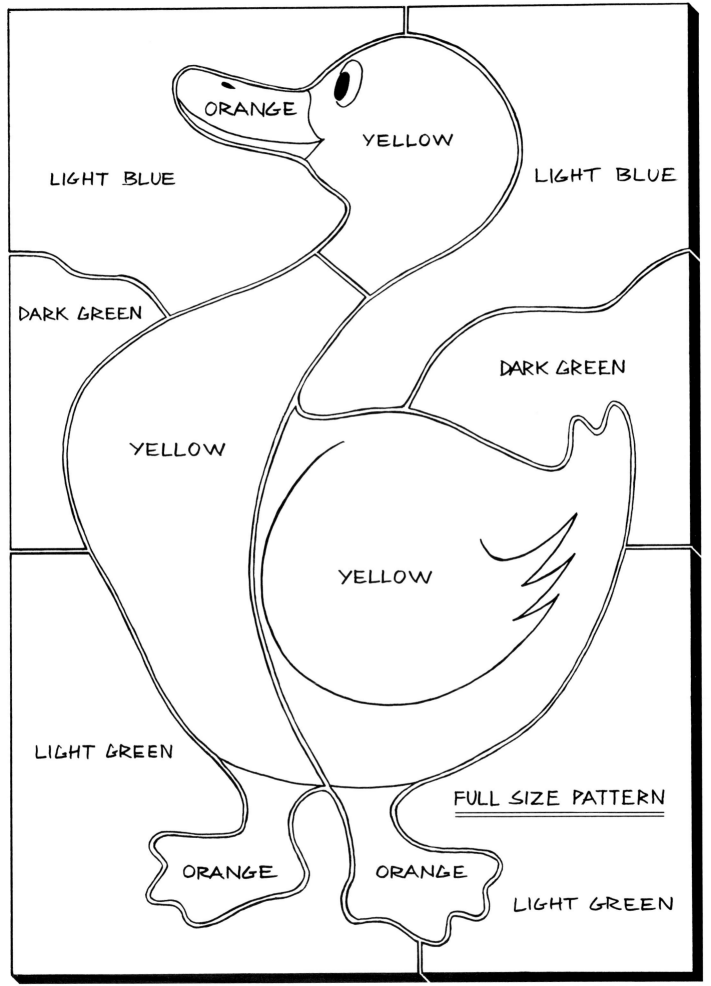

Country Occasional Table

We've always liked basic pine projects that can be constructed with standard ¾ in. thick pine boards using everyday workshop tools.

This small table is one of the nicest "occasional table" designs that we've seen in sometime. While you could edge-glue narrower material to obtain the wide stock required, mill-finished, glued-up wide pine is also available at most lumber yards.

There are only five major parts to the table: the two sides (A), the stretcher (B), the shelf (C), and the top (D). Although the bill of materials provides the *final* overall dimensions for all parts, it is usually best to rough cut your material slightly oversize. In the case of the sides and shelf, this extra material is especially important since the final crosscuts establishing the length of these parts are at an angle.

Bill of Materials
(all dimensions actual)

Part	Description	Size	No. Req'd.
A	Side	¾ × 11 × 21⅜	2
B	Stretcher	¾ × 3 × 18¼	1
C	Shelf	¾ × 8 × 15	1
D	Top	¾ × 14 × 22	1

To make the sides, first rip material to the approximate overall width, and with the table saw blade inclined at 5 degrees, crosscut both ends to obtain the final length. Then use a tapering fixture to establish the tapers on both sides. While the notch to accept the stretcher tenon can be cut by standing the sides on end and using the dado head, we made the notch by hand using a guide block and chisel, as illustrated in the detail. Lay out and saber saw the 11 in. radius curves along both leg tapers, and lay out and cut the bottom end arch and heart profiles using the 1 in. square grid patterns as a guide.

The stretcher can be made next. Note that the lower ends that butt to the sides must reflect the 5-degree angle of the sides. As shown, the ½ in. diameter dowel holes are slightly offset to tension the sides when the dowels are inserted.

The shelf ends must also reflect the 5-degree angle of the sides, and are crosscut on the table saw with the blade set at 5 degrees.

The top is simply cut to size, and radiused on the top and bottom edges as shown.

The top is fastened to the sides and the sides joined to the shelf with ⅜ in. dowel pins as illustrated in the exploded view. You could make special jigs to drill these holes, although a small hand held drill and a good eye should be sufficiently accurate. The dowels are drilled for an approximate 10-degree angle to provide some additional mechanical strength.

Because of the 5-degree side angle, you may find that it is difficult to keep the shelf from slipping out of place when bar clamps are applied across the two sides. You could clamp stopblocks in place to prevent such movement, but we prefer to insert several small brads in the shelf ends and then nip the heads so that about ⅛ in. remains. The brads prevent any slippage. If you are using pre-glued wide pine stock, note that this material typically measures only ¹¹⁄₁₆ in. thick. Substitute ⁵⁄₁₆ in. diameter dowels for the ⅜ in. diameter dowels to avoid splitting out the ¹¹⁄₁₆ in. material. If you use brads in the shelf ends, be sure to keep them clear of the dowel locations.

If everything in this project doesn't turn out perfectly, don't worry. With a piece that is going to be distressed, a little rough work adds to the look of authenticity.

We distressed our table by gouging it with a chisel. The key when distressing is to know when to stop. Not enough distress marks, and they look like defects; too many and the piece becomes busy. After distressing, we sanded, stained and then oiled the table with a good quality penetrating oil.

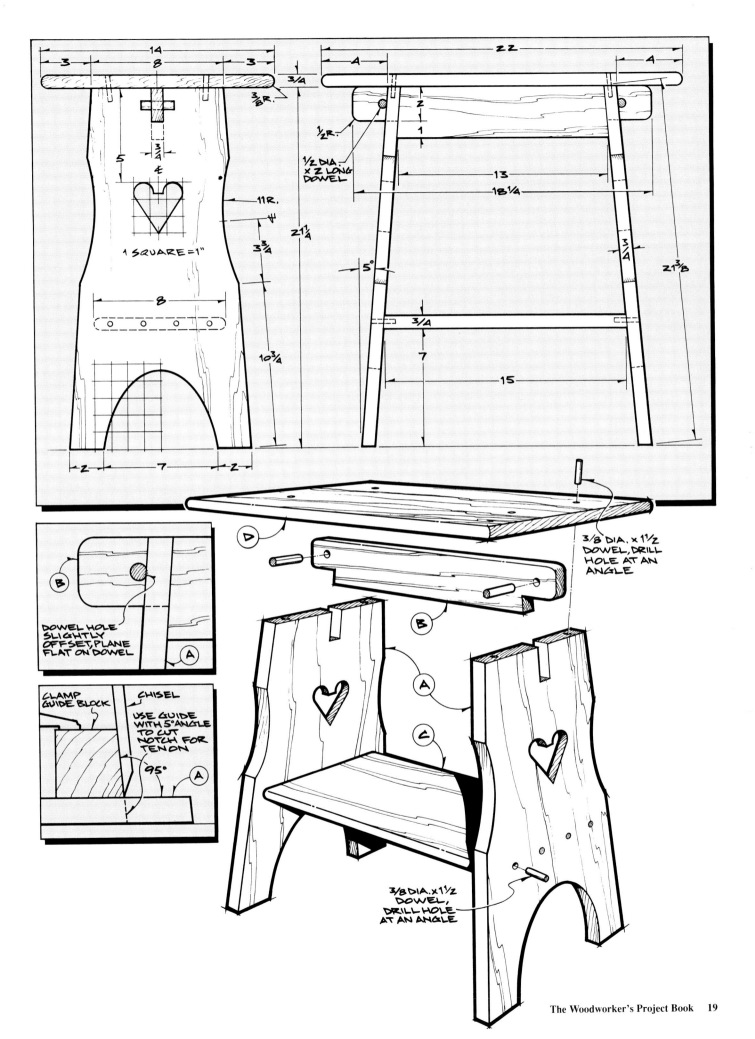

Early American Candlesticks

Add an elegant look to your dining table with this handsome pair of Early American style candlesticks. We made ours from cherry, but mahogany and walnut are also good choices.

You'll need to glue up a turning blank for each candlestick. A turning blank is made up of three pieces of stock, each piece measuring 1¾ in. thick by 5¼ in. wide by 7½ in. long. Face-glue the three pieces together, then clamp firmly.

Dress the glued-up stock to 5 in. square, then trim the four corners to create an eight-sided (octagonal) blank. Next, mount the blank to the lathe plate. We used a 3 in. diameter faceplate and secured it to the blank with 1¼ in. long by no. 12 flathead wood screws.

Mount the faceplate to the lathe, then position the tool rest for the initial turning cuts. In its proper position, the edge of the tool rest should be parallel to the blank and almost touching it. Also, it should be just above the center line of the blank. Before starting the lathe, make sure there is adequate clearance between the blank and the tool rest. You can check for this by hand-turning the blank at least one revolution.

Next, using a gouge, turn the blank into the shape of a cylinder. Keep in mind that the octagonal blank may not be perfectly balanced as it rotates, so it's important to do this initial turning with the lathe at its slowest speed. Once the blank has been shaped into a cylinder, the lathe speed is increased for the remaining operations. The increased speed results in faster and smoother cuts.

Now, reposition the tool rest so that it is parallel to the end of the blank. Its height should be adjusted so the cutting edge of your chisel is at the blank's center line. Lock the tool rest in place, then use the roundnose chisel to flatten and smooth the end of the stock.

The candlestick can now be turned as shown in Figs. 1 through 5. A summary of the procedure is as follows:

Figure 1: Return the tool rest to its original position (parallel to the length of the blank and slightly above the center line) and, with the tool rest locked in place, use a parting tool to establish the 6⅛ in. length shown. Make the parting tool cut about ⅜ in. wide by ⅜ in. deep.

Figure 2: Use the parting tool to make a cut establishing the 5¹/₁₆ in. length and the 2⅜ in. diameter. Once the parting cut has been made, use the roundnose chisel to turn down the remainder of the cylinder to 2⅜ in. diameter.

Figure 3: With the skew chisel, round the end and base as shown.

Figures 4 and 5: Turn as shown using the gouge, roundnose, and skew as needed to create the various profiles. A diamond point chisel may be helpful for cleaning up the base profile.

Once the turning has been completed, the piece is sanded while still on the lathe. Start with 100 grit sandpaper, then follow with 150 and 220 grit. For an exceptionally smooth surface, add a final sanding with 400 grit.

The parting tool is used to cut off the blank. Make the parting cut at the bottom of the candlestick base; however, don't cut all the way through. Instead, allow about ½ in. diameter to remain. Remove the faceplate from the lathe, then unscrew the faceplate from the blank and use a hand saw to cut through the remaining ½ in. diameter.

Sand the bottom of the candlestick until it is flat. Once the bottom is flat, use the drill press with a Forstner bit to bore a ⅞ in. diameter by ⁹/₁₆ in. deep hole for the candle cup ferrule. If not available locally, the ferrule can be ordered from The Woodworkers' Store, 21801 Industrial Blvd., Rogers, MN 55374 (specify part no. 21766).

For a final finish we applied three coats of tung oil, rubbing out between coats with 0000 steel wool.

The ferrule can now be added. Protect the top edge of the ferrule with a block of scrap wood, then use a hammer to tap it in place. The addition of a candle completes the project.

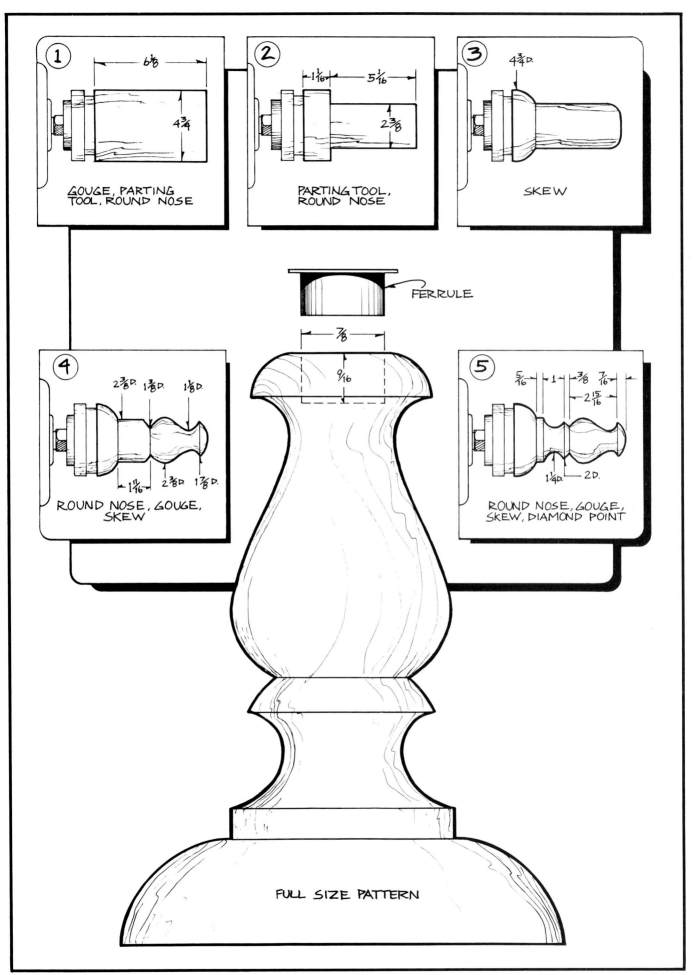

Arrow Wall Decoration

This attractive wall ornament mounted over a mantel will lend an 18th-century touch to your living room.

The antique pine weather vane that inspired the design no doubt swung toward the northeast on many a windy March day. But a cozy spot above your hearth will suit the piece just as well.

We made ours from ¾ in. mahogany and finished it with penetrating oil. The original piece is featured in *The Pine Furniture of Early New England* by Russell Hawes Kettell.

Start with a blank 6 in. wide and 43 in. long. Lay out the grid pattern on paper and glue the full-size drawing to the stock with rubber cement.

First cut the outside profile with a ⅛ in. blade in the band saw. Stay outside the lines. Clean up the cuts with several sizes of small drum sanders.

Start shaping the inside by drilling five holes, one at the center of the half

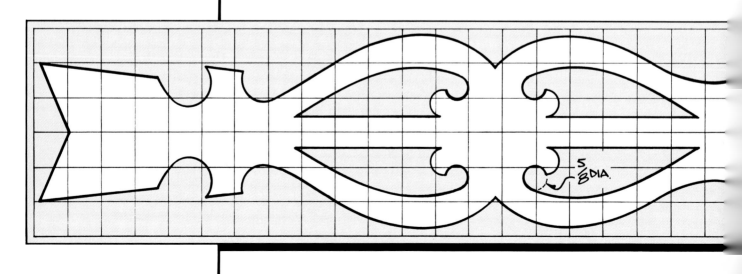

circles and four at the tight radii at the top of the hearts. Drill the holes with $\frac{5}{8}$ and $\frac{3}{4}$ in. Forstner bits as shown.

Because of the arrow size, a scroll saw won't work on the inside cuts, so use a hand-held jigsaw with a fine blade or a coping saw. To start your cuts, use your drill holes and make several more in strategic spots of the waste material. Cut the gentle curves and straight sections first, letting the waste drop out. When you cut the tight curves, most of the material will be gone, allowing free movement of the saw blade.

There is just no easy way to clean up the inside cuts. We used a series of small files: triangular, half-round, flat, $\frac{1}{2}$ in. diameter round and a small jewelers' file. Hand sand and apply two coats of penetrating oil.

To hang the piece, cut two $\frac{1}{4}$ in. diameter holes in the back and mount on picture hangers.

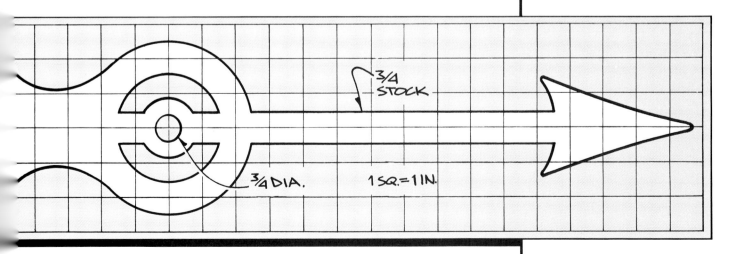

Three-Drawer Country Wall Box

This attractive wall box can be used practically anywhere for storage of small items or knickknacks. Additionally, its drawers hold three by five recipe cards, making it an ideal kitchen accessory.

Adapted from traditional Early American style wall boxes, our version includes a base so it can also be located on a shelf or countertop. While our box is crafted from pine, other woods could be used as well. Oak, cherry, or walnut come to mind as good choices.

You will note that the box itself is constructed of ½ in. thick stock, while ¼ in. stock is used for the drawer bodies. The drawer bottoms are ¼ in. plywood, and the drawer faces are ½ in. stock. While we surface all our project stock ourselves to obtain the desired thickness, readers buying nominal pre-surfaced material should note that actual thicknesses may be slightly under the specified dimension. For example, nominal ½ in. stock may actually measure only ⁷⁄₁₆ in. Adjust dado width and other dimensions as needed if you use thinner stock.

Start by cutting the ½ in. pine for the sides (A), back (B), base (C) and shelves (D). Next lay out the grid patterns and cut the scrolls on the back and sides as shown. Mount the dado blade in a table or radial-arm saw and cut the through dadoes in the sides for the shelves. Cut the bead on the base with a ⅜ in. round-over bit in the router.

The ⅛ in. diameter slotted screw holes in the back allow it to expand or contract independently of the box. The two ³⁄₁₆ in. diameter mounting holes are for attaching the box to the wall.

Sand before assembling and use glue sparingly to prevent it from squeezing out when the piece is clamped. To assemble, first fit the four shelves in the dado joints, then attach the back with 1 in. long by number 6 flathead wood screws. Square up the piece before tightening the screws.

Drill and peg the sides into the shelves and fasten the base with glue and a clamp.

When making the drawers, remember to size them to the actual openings, keeping them slightly smaller so they slide easily. If your box opening measures 5⅝ in. across as shown, make the drawer fronts (E) and backs (G) ¹⁄₁₆ in. shorter than indicated.

A ⅛ by ¼ in. groove in the drawer sides and front houses the drawer bottom (H). When cutting the drawer parts and the grooves for the bottoms, remember to set the saw once for each operation. This insures the parts are the same size and will fit neatly together. The drawer boxes are assembled with finish nails.

If you want, you can cut all the drawer and box parts at the same time, just leaving drawer parts slightly long (there's plenty of built-in clearance in the width). You can shave them to the proper size just before assembly. That way you can cut all the parts, remove the saw blade and cut all the grooves with the dado blade.

To cut the drawer faces (I), set the table saw blade at 65 degrees and use a high fence for stability as you pass the workpieces past the blade. For even greater stability, use a finger board. Cut the end-grain sides first to prevent splintering from marring the finished face. Use a push stick for the operation as the pieces are rather small. Fasten the faces to the drawers with ½ in. long screws.

We finished the piece with Minwax Special Walnut stain, followed by an application of satin polyurethane.

Traditional porcelain drawer pulls (J) complete the project.

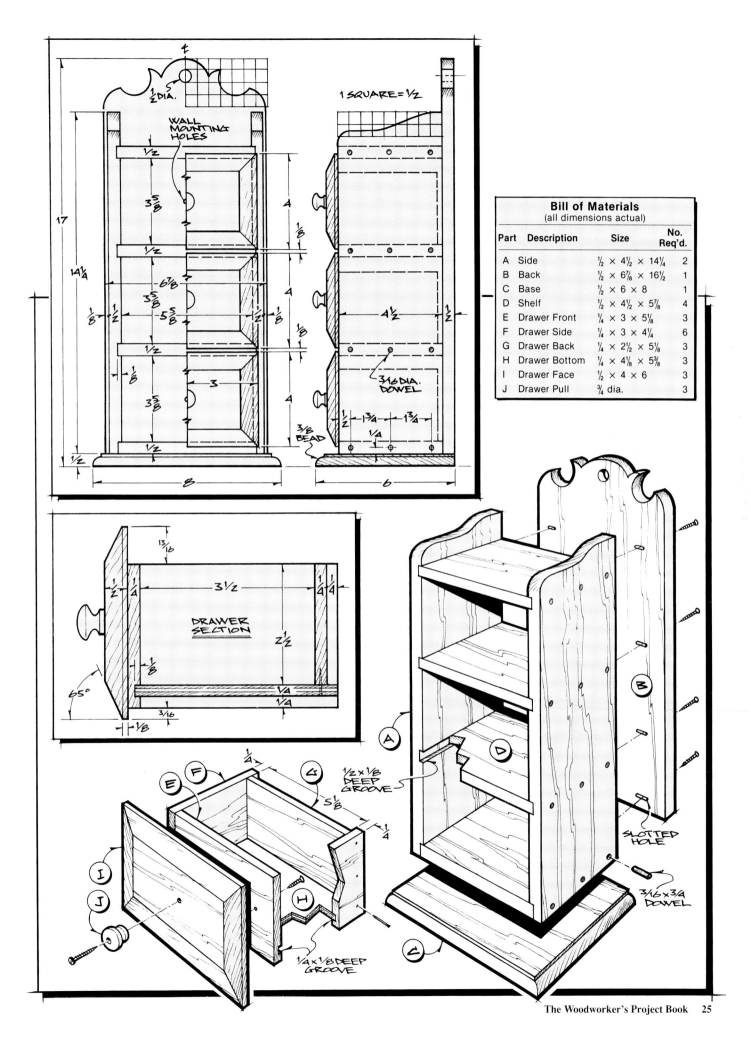

Hanging this simple cabinet beside the door may save moments of frantic searching when you're already late for work. And hung out of reach, it may also prevent your keys from walking away to a toddler's favorite corner.

The piece is made from pine, available at any lumber yard, and finished with Minwax Golden Oak stain. For the ½ in. thick door you may have to mill down a ¾ in. pine board, using a hand plane or jointer.

Start by cutting the pieces to size, leaving a little extra on the width of the door (E) to allow for some shrinkage.

Join the sides (A) to the top and bottom (B) with ¼ in. diameter by 1 in. long blind dowels. Lay out the dowel locations with a square. Note that the

Key Cabinet

dowels are placed ⅜ in. from the front edge, but ⅝ in. from the rear edge to account for the ¼ in. deep rabbet housing the back.

Before marking, orient the sides, top and bottom the way you want them for final assembly.

After marking the holes, clamp the piece to the table of the drill press, and drill one side of each joint with a Forstner or brad-point bit. If you don't have a drill press, just do your best to hit 90 degrees. The soft pine will forgive a slight skew. You can then find the opposite holes using your marks. Or, to prevent errors from creeping in, use dowel centers.

You will cut the dowels 1 in. long, so make the holes slightly deeper than ½ in. to allow the glue some room and prevent the dowels from holding the joint open.

Glue up the box using two clamps. Apply glue to the dowels only since the end grain won't contribute to strength.

After the glue dries, remove the clamps and cut the ¼ in. by ⅜ in. rabbet

Bill of Materials
(all dimensions actual)

Part	Description	Size	No. Req'd.
A	Side	¾ × 2 × 9½	2
B	Top/Bottom	¾ × 2 × 4½	2
C	Back	¼ × 5¼ × 8¾	1
D	Mounting Block	½ × ¾ × 4½	2
E	Door	½ × 6 × 10¾	1
F	Hinge	1 × 1 brass	2
G	Knob	⅝ dia. brass	1
H	Cuphook	1 in. as shown	6
I	Magnetic Catch	as shown	1

for the back (C). We used the router table with a ⅜ in. bearing-guided router bit set to a ¼ in. depth.

Transfer the door pattern to the milled pine. Cut the curves with a fine blade in the scroll saw or hand-held jigsaw. Sand the outside edge, then set the ¼ in. cove into the edge of the door with the router table as shown.

Next, set the back into the box. Either square up the rabbet with a chisel or put radii at the corners of the ¼ in. plywood. The radius must match the rabbet cut into the box by the router. Our ⅜ in. bit had a 1¼ in. diameter and a ⅝ in. radius, but your bit may be different. Just measure the diameter.

To ease clamping of the mounting blocks (D), set the back into its rabbet and mark their locations. Then remove the back and clamp the mounting blocks in place. Use a thin line of glue on both block and back. While the glue is drying, mortise out the door and sides for the hinges (F). Finally, drill and countersink holes in the back: ten to join the back to the cabinet and two for mounting on the wall. Use no. 6 by ½ in. flathead wood screws for fastening, and no. 8 by 1½ in. flathead wood screws for mounting.

Sand the piece before assembling. Mount the magnetic catch (I) before the cuphooks (H). The hinges, knob (G), and catch are all available from Constantine's, 2050 Eastchester Rd., Bronx, NY 10461. The cup hooks are available at any hardware store.

We used a brass knob; however, if you prefer, a wood knob can be turned to the dimensions shown in the detail.

26 The Woodworker's Project Book

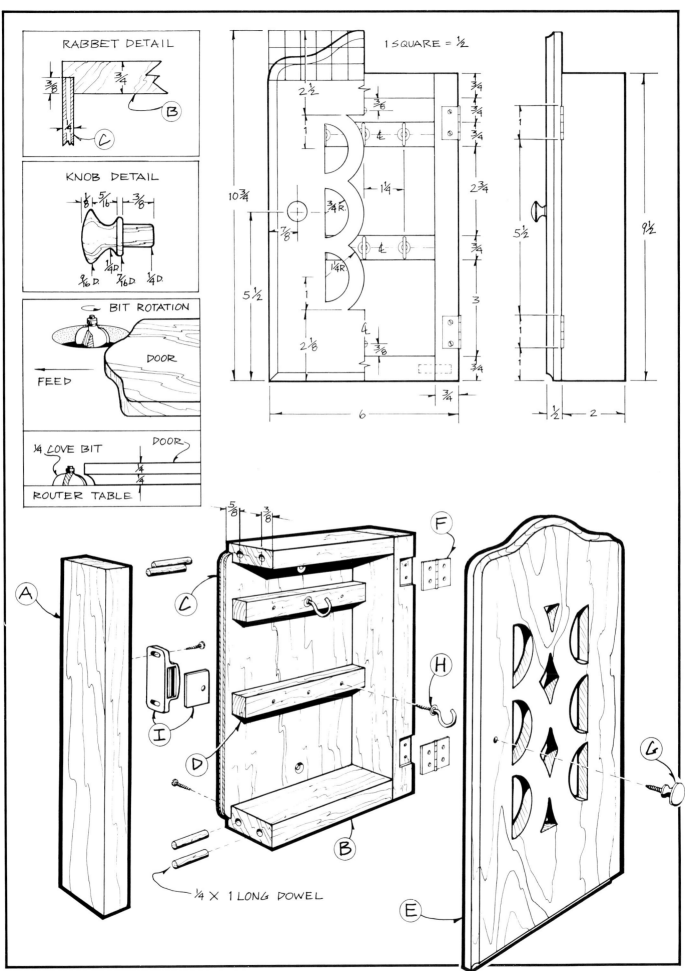

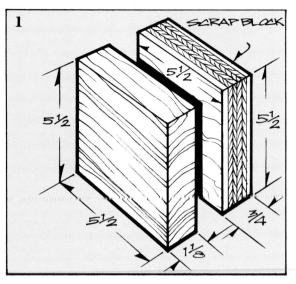

Figure 1: Cut the block for the box and a piece of scrap plywood to the dimensions shown.

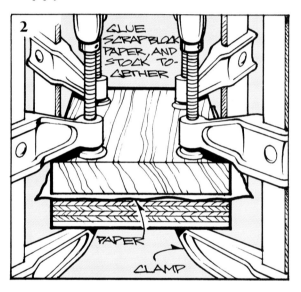

Figure 2: Glue and clamp the two blocks together using brown kraft paper in between.

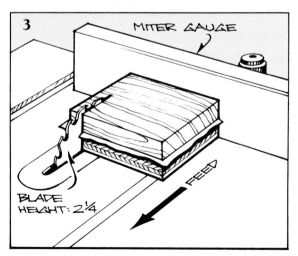

Figure 3: After the glue dries, set the table saw blade to 2¼ in. and, using the miter gauge, crosscut the block to 5 in.

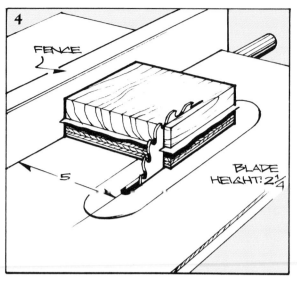

Figure 4: Rip the block to 5 in.

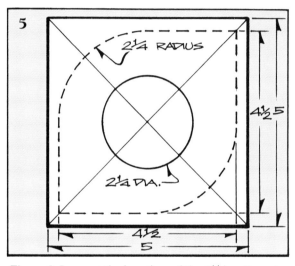

Figure 5: Lay out the location of the 2¼ in. diameter hole and also the outside profile of the box.

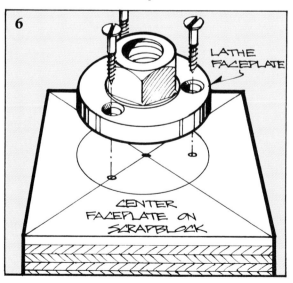

Figure 6: Find the center of the scrap side of the block, inscribe a circle and fasten your faceplate.

CONTEMPORARY BOX

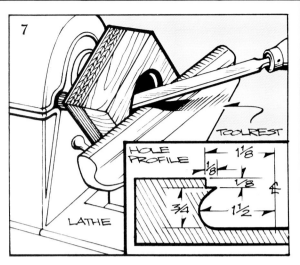

Figure 7: Set the lathe tool rest about ¼ in. below the center of the work and rough out the opening. Setting the tool rest low will prevent a nub from forming at the center of the piece. Use a spear-point chisel for the ⅛ in. by ⅛ in. lip and a roundnose for the inside cove. Be careful to angle your tools from the rear of the lathe toward the front.

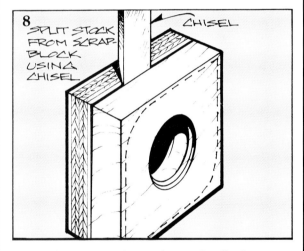

Figure 8: Remove the turning and split the scrap block off with a chisel, working the tool back and forth until the piece starts to separate.

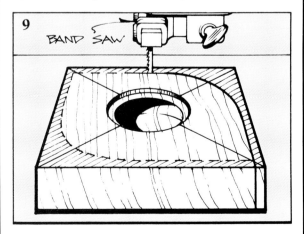

Figure 9: Band saw the outside profile.

(continued on next page)

This small box, designed by California woodworker Robert Leung, makes a useful addition to any desk. We find it's just the right size for paper clips or stamps.

Leung used pau ferro (Santos rosewood) for the body, and bird's-eye maple and East Indian rosewood for the lid. He also makes similar boxes in various sizes and different combinations of exotic and domestic hardwoods. A mail-order source for rare or unusual hardwoods is Berea Hardwoods Co., 125 Jacqueline Drive, Berea, OH 44017.

With the angles and inside turning, the project may appear difficult, but any moderately experienced woodworker can successfully complete the box by following this step-by-step procedure. We suggest cutting the box angle, a 10-degree slope toward the bottom, with a disc sander after cutting the profile with a band saw. The inside turning is easily done with roundnose and spear-point chisels.

The inside dimensions aren't critical, so don't worry if you don't exactly match the plans. For a snug fit be sure to cut the circle for the lid after making the opening.

To begin you need a block of ⅝ in. pau ferro (or another hardwood) 5½ in. square, a scrap block the same size, a piece of maple at least 2¾ in. square and a strip of rosewood ¾ in. by ¾ in. by 4 in. Then just follow our step-by-step procedure that begins on the facing page.

CONTEMPORARY BOX

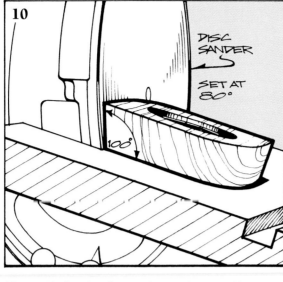

Figure 10: Set the disc sander 10 degrees off square and grind the bevel to slope toward the bottom.

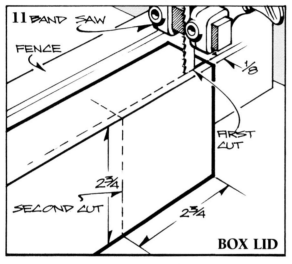

Figure 11: To get stock for the lid, resaw a piece of maple to the dimensions shown, using a ½ in. or wider blade. Clamp the ⅛ in. thick lid stock flat to let it acclimate before cutting the lid to shape. If the lid stock is not allowed to acclimate, it may warp and possibly shrink out of round over time.

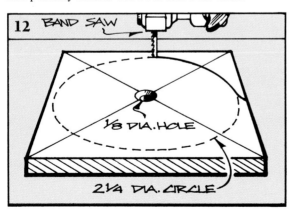

Figure 12: Drill a ⅛ in. shank hole in the center of the lid stock and countersink the hole. Then lay out and cut the circle on the band saw.

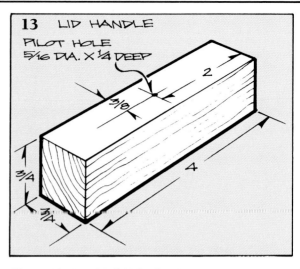

Figure 13: Cut the lid handle block to size and drill the pilot hole. Lay out the curves with trammel points or a compass.

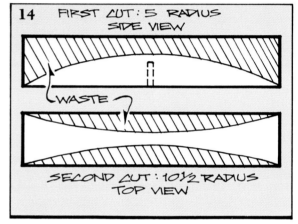

Figure 14: Cut the 5 in. radius in the handle, being careful to cut the side away from the pilot hole, and then cut the two 10½ in. radii on the sides. Stay outside the lines.

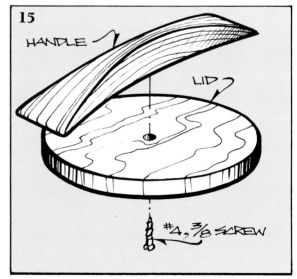

Figure 15: Clean up the cuts with a drum sander, thoroughly hand sand all parts, and assemble. Finish with a clear penetrating oil.

Contemporary Glass-top Table

A narrow band of red dyed veneer on each leg adds a striking detail to this handsome glass-top table. Maple, with it's naturally light color, contrasts nicely with the veneer, so we used it for all wood parts. You'll need 1¾ in. thick stock for the legs (A) and ¾ in. thick stock for the end stretchers (B) and front and back stretchers (C).

Begin by cross-cutting stock for the legs and stretchers to approximate length, allowing about 1 in. extra for each part. Next, rip each of the parts to approximate width, adding an extra ¼ in. or so for each cut.

Use the jointer to plane (joint) one of the ripped edges of each part. Make the jointer cuts in at least two passes, with the final pass removing no more than about ¹⁄₃₂ in. of stock. A light cut like this produces a smooth finish that requires little sanding later on.

The stock can now be ripped to final width. To get a smooth finish on the final cut, we like to set the rip fence to allow an extra ¹⁄₃₂ in. on the width. After ripping, the jointer is used to remove the added ¹⁄₃₂ in.

Now, crosscut the stretchers to final length. The cuts must be square here, so it's a good idea to first check your table saw setup.

The fixtures (Fig. 4) will help to insure accuracy when shaping the stretchers. They also eliminate much of the tedious sanding of the curves. However, if you prefer not to make the fixtures, just lay out the stretcher profile on each part and cut it out with the band saw, then sand the edges smooth.

To make the fixtures, cut the plywood to size and mark the center points of the ½ in. radius curves, then use a 1 in. dia. bit to bore each hole. With the table saw blade set to a height of ¾ in., use the miter gauge to pass the plywood through the blade to establish cuts no. 1, 2, 3, and 4 as shown in Fig. 1 on page 33. Note that the cuts are made tangent to the 1 in. diameters. Now, mark the location of cuts no. 5, 6, and 7, again making sure that the cut lines are tangent. Use the band saw to make these cuts, taking care to stay about ¹⁄₁₆ in. on the waste side of the line. A sanding block is used to sand the band saw cuts to the marked lines.

To complete the fixtures, cut ¾ in. square by 2¼ in. long blocks. Screw them in place (first bore pilot holes) as shown in Fig. 4, then drive a pair of 1½ in. by no. 8 screws through each end block. The screws hold the stock in place when the trim cuts are made later.

The fixtures can now be used to trim the stretchers to final shape. Begin by placing the top face (the face without the blocks) of the short (15 in.) fixture on one of the end stretchers, making sure that all four edges of both parts are flush. Use a sharp pencil to scribe each curve, then cut them out with the band saw. When making the band saw cuts, it's important to stay about ¹⁄₁₆ in. on the waste side of the marked line.

Place the end stretcher into the fixture and tighten the two end screws. Equip the router table with a laminate

The Woodworker's Project Book 31

flush trim bit that has a flute length of at least 1 in. and, as shown in Fig. 2, set the bit to a height that allows the ball-bearing to run against the edge of the fixture. Trim each of the band saw cuts as shown, taking care to keep your hands a safe distance away from the bit. This same procedure is followed for the remaining stretchers.

Now lay out and mark the location of the mortises on the stretchers and legs and cut them out with a sharp chisel. Since the legs were originally cut with 1 in. added to the length, you need to locate these mortises $\frac{1}{2}$ in. down from the top end. Later, when each leg is cut to final length, the end will be trimmed so that the mortise location is $\frac{3}{16}$ in. below the top.

Make the four splines from solid stock. For maximum strength, the grain direction must run as shown in the exploded view.

Next, cut the $\frac{1}{2}$ in. radius coves in each leg. We used the router table and a $\frac{1}{2}$ in. radius bearing-guided cove bit to do the work. To get a smooth cut with a minimum of strain on the motor, it's best to make each cut in four passes. For the first pass, set the bit to make a $\frac{1}{8}$ in. deep cut, then raise the bit $\frac{1}{8}$ in. for each of the remaining three passes.

Now that the leg coves have been cut, the dyed veneer (D) can be applied. We used the $\frac{1}{40}$ in. thick Cherry Red dyed veneer sold by Constantine, 2050 Eastchester Road, Bronx, NY 10461. It's their part no. DV307 and it comes in 36 in. long sheets that run 6 in. to 10 in. wide. Keep in mind that there is a minimum order requirement of three square feet.

Use a razor knife to cut four strips of the veneer, each strip measuring 1 in. wide and at least $16\frac{1}{2}$ in. long. This provides more than enough material and the excess will be trimmed later.

To support the leg during glue-up of the veneer, we made a V-groove base and five V-groove clamp blocks (Fig. 3). Also, as shown, we planed a flat

Bill of Materials
(all dimensions actual)

Part	Description	Size	No. Req'd
A	Leg	$1\frac{3}{4} \times 1\frac{3}{4} \times 15\frac{1}{2}$	4
B	End Stretcher	$\frac{3}{4} \times 2 \times 13\frac{1}{2}$	2
C	Front/Back Stretcher	$\frac{3}{4} \times 2 \times 41\frac{1}{2}$	2
D	Veneer	$\frac{1}{40}$ in. thick	as req'd.
E	Glass Top	$\frac{3}{8} \times 18 \times 46$	1
F	Glide	$\frac{7}{8}$ dia.	4

edge on two pieces of 1 in. diameter dowel stock. The dowel is used to press the veneer in place, and the flat edge provides a good bearing surface for the clamp. A coat of paste wax added to the dowel prevents it from sticking to the veneer.

Working with one leg at a time, apply a thin coat of white or yellow glue to both coves and to the veneer. Now use the dowel stock to press the veneer into the cove, then add the five clamps, evenly spaced, to apply pressure. Allow the glue to dry thoroughly.

After the veneer has been applied to all four legs, use a razor knife to trim the excess veneer along the length of each piece. However, don't trim it perfectly flush with the surface of the leg. It's best to allow about $\frac{1}{16}$ in. to remain and then use a sanding block to sand it flush. Any excess glue can be cleaned up as you sand.

Next, crosscut the legs to final length. Cut the top of the leg first, establishing the $\frac{3}{16}$ in. distance to the mortises, then measure the $15\frac{1}{2}$ in. overall length and make the bottom cut. To complete work on the legs, drill and counterbore the bottom end (Fig. 6) for the glides (F) which will be added later on.

Final sand all parts finishing with 220 grit paper. At this point, dry assemble all the legs and stretchers to check for proper fit-up. If everything looks okay, add a thin coat of paste wax to the areas around the joints where glue squeeze-out is likely to occur. Since the glue won't stick to the wax, it makes clean-up much easier later on.

The table can now be assembled. First, though, make four corner clamp blocks (Fig. 5) to protect the legs from clamp damage. Apply glue to each of the mortises and splines, then assemble all eight parts and clamp firmly. To make sure the top surface of the table is perfectly flat, it's important to assemble the table upside-down on a flat surface. We used the top of our table saw. If this

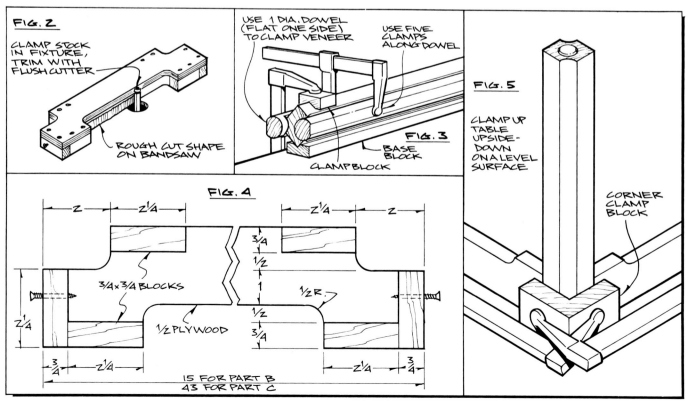

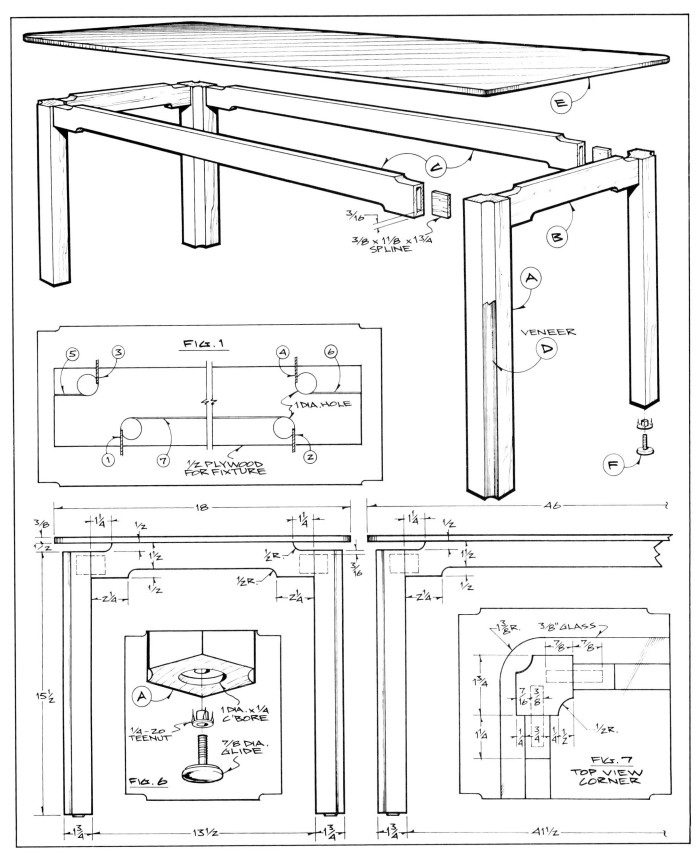

piece is built with a top surface that is not perfectly flat, the glass top will tend to rock, so take the time here to make sure everything is right. Once you are satisfied all is okay, set the project aside to dry.

Once dry, a sharp chisel will clean up any glue squeeze-out, and any remaining wax can be removed with acetone. Use epoxy to glue the teenuts into the leg bottoms and, when the glue dries, add the levelers. For a final finish, we sprayed on two coats of Deft's Gloss Clear Wood Finish, followed by a thorough rub down of all surfaces with 0000 steel wool.

The 3/8 in. thick glass top (E) can be ordered at any glass shop. We specified a "factory polished edge" in order to get a highly smoothed edge all around. Note, as shown in the detail, that the corners have a 1 3/8 in. radius. No special means of fastening is necessary, the glass simply rests on the stretchers.

Shaker Carrier

This carrier, by craftsman Joe Robson of Trumansburg, New York, was a juried selection in the Shaker Workmanship 1987 exhibit at Hancock Shaker Village in Hancock, Massachusetts.

Robson tells us that the carrier is a hybrid design, patterned generally after Shaker carriers of a similar style. Coincidentally, an original Shaker piece nearly identical in every respect to his was sold at a New Hampshire auction in 1984 for over $4,000.

The carrier is crafted in black cherry with a white ash bentwood handle, and features a simple dovetail construction, with a raised profile bottom panel in a groove housing. If you are unable to resaw to obtain the 1/4 in. thick stock for the sides, ends, and bottom, 1/4 in. thick cherry stock surfaced both sides can be ordered from Craftsman Wood Service, 1735 West Cortland Court, Addison, IL 60101.

If you have never tried hand-cutting dovetails, this is an ideal project to get started. As for proper terminology, the segments cut on the ends of parts A are called the dovetails, presumably because they look like the tail of a dove. The pins are the corresponding parts on the ends (B) that fit in between the dovetails on parts A. Begin by laying out the three dovetails on each end of part A. Ideally, the length of the dovetail should be equal to the thickness of part B (1/4 in.), plus about 1/32 in. When the pins are cut, they too will be slightly longer (1/32 in.) than the thickness of the sides. Later, when the joint is assembled, the dovetails and the pins will both stand a little proud and will be sanded flush with the carrier sides.

Lay out the dovetails very carefully using a hard, sharp pencil. Once the tails have been laid out, mark the waste material between with an "X" to avoid confusion. In addition to scribing the tail layout on the face of the board, carry the lines across the end grain. Now clamp the side in a vise and use a fine-tooth dovetail saw to make the angled cuts that will establish the tails. Cut just on the waste side of the line, grazing but not removing it. Bring the cuts almost — but not quite — to the scribed bottom line. A coping saw is now used to cut across the grain and remove the waste. Take the side from the vise, clamp it flat on the workbench over a scrap board, and use a sharp chisel to dress the sides and bottoms of the cutouts.

The pins on the carrier ends should be laid out and marked using the finished dovetails as a template. To do this, clamp the end vertically in a vise, lay the dovetailed side in position on the end, and trace the dovetails with a sharp pencil or X-acto knife. Use a small square to carry the scribed lines onto the face of the board, remembering to add the extra 1/32 in. as noted earlier.

Mark the waste portion between the pins with an "X", then cut out using the same technique as for the dovetails. A well-fitted dovetail should fit together with only light tapping from a mallet and scrap block. If needed, trim further with a sharp chisel.

Now cut the 1/8 in. deep by 3/16 in. wide groove that houses the bottom, using a 3/16 in. straight bit in the router table. Note that the groove must be stopped on the sides. Robson cautions the need for extra care here to avoid chipping out the dovetail. The raised panel effect on the bottom can be made on the table saw by passing the panel across the blade on edge, just kissing the blade, to create the 1/16 in. deep rabbet all around. Note that the panel dimensions are sized slightly less than the actual groove-to-groove distance to accommodate possible expansion/contraction.

Bill of Materials
(all dimensions actual)

Part	Description	Size	No. Req'd.
A	Side	1/4 × 3 3/8 × 11 3/8	2
B	End	1/4 × 3 3/8 × 6	2
C	Bottom Panel	1/4 × 5 5/8 × 11 1/16	1
D	Handle	1/8 × 7/8 × 26	1

Assemble the sides and ends with the bottom in place, taking care to apply glue to all mating surfaces of the dovetails and pins. No glue is needed on the bottom, except for a dab at each center end to equalize movement.

Robson notes that the handle (D) should be made only from quartersawn ash with very straight grain. Cut the handle stock to length, width and thickness, and shape the taper. The portion of the handle that is bent is simply steamed with live steam, using a kettle with a spout. When pliable, clamp immediately in place on the carrier. If needed, lift the box from the center of the handle to center the curve.

Robson uses a traditional Shaker finish, made by shredding three ounces of beeswax into a pint of turpentine to dissolve it, and then mixing in one pint of boiled linseed oil. The mixture is applied liberally with a rag, then wiped dry once it starts to tack. Robson adds that the traditional Shaker method required that the finish be applied "once a day for 10 days, once a week for 10 weeks, and once a year thereafter." Cherry will darken naturally over time, a process that can be hastened by exposure to direct sunlight.

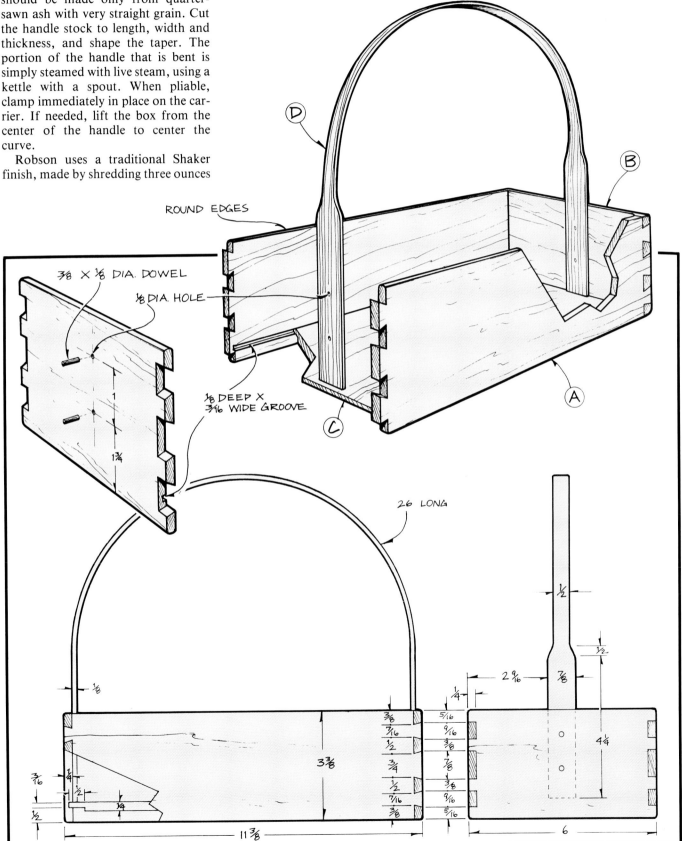

HUNT TABLE

Tradition has it that the Hunt Table was a long table positioned near an entrance where hunters laid out their game upon returning home. Our modern version of this classic table is somewhat smaller than the early tables, and will function nicely as a sofa table or hall table. The 16 in. width fits well in hallways and other places where large tables create an obstacle.

Our table is crafted in cherry, an elegant hardwood that darkens to a deep rich color over time. Oak, walnut, or a figured maple would be other choices you might want to consider. If you opt for oak, however, be prepared for a little additional work, since oak is a difficult wood to turn.

Start by getting out stock for the top (A). Unless you have saved a very special wide board for just such a project, you'll have to edge-glue two 8 in. boards to get the 16 in. top width. While you're waiting for the top to dry, get out sufficient $\frac{3}{4}$ in. stock for parts B, C, D, E, G, and H. Also, rip and joint $\frac{8}{4}$ stock to $1\frac{3}{4}$ in. square for the four leg (F) turning blanks.

Now lay out the apron and stretcher mortise locations on the leg blanks. Take note that the various mortises are cut before the legs are turned. We cut the leg blanks long enough to include at least 1 in. of waste on each end for mounting in the lathe. However, you could simply cut the legs to their final length before turning. Just be careful when turning the bottom end not to let the cutting tool slip off and into the lathe center.

We use a plunge router and a $\frac{1}{2}$ in. diameter straight cutter to establish the mortises. Use the edge-guide and set up stopblocks to establish the mortise lengths. Cut the $1\frac{1}{16}$ in. mortise depth in eight or nine passes, each pass removing about $\frac{1}{8}$ in. of stock.

Next, mount the legs in the lathe and turn to the profile shown. As you might note, vase shapes are essentially elongated combinations of beads and coves. Final sanding of the legs should also be done while mounted in the lathe.

After cutting the tenons on the ends of the aprons and stretchers, use a $\frac{3}{8}$ in. beading bit to establish the bead detail

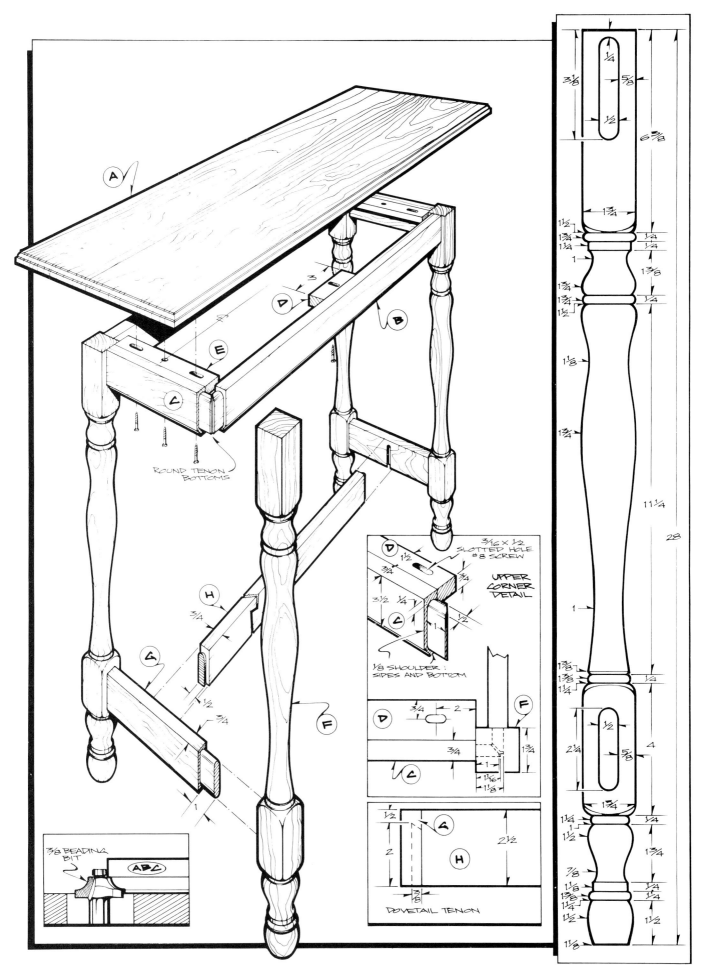

on the aprons and the top (see detail). Rout the slotted screw holes into the cleats that will allow the top to move freely in relation to the frame.

The dovetail groove in the end stretchers is cut with the router and a ½ in. dovetail bit, using a board fence clamped across the stretcher as a guide for the router. Stop the groove about ½ in. from the top edge. To cut the dovetail on the ends of the center stretcher, you'll need to use the router table. Stand the stretcher on end, and clamp a 12 in. long guide board to the stretcher to guide off the top of the router table fence, as illustrated in the detail. The guide board eliminates any tendency of the stretcher to rock or tip as you are passing it across the cutter. Use a chisel to trim back and undercut the dovetail tenon at the top, enabling it to fully seat into the dovetail end stretcher groove, so that when assembled both stretchers are flush.

To assemble the table frame, first complete the two leg, end stretcher, and end apron assemblies; then add the side aprons, turn that assembly upside down, and tap the center stretcher into place. Apply glue to the blind half of the dovetail groove and the bottom half of the dovetail to facilitate assembly. If you load glue the full length of the dovetail, it may lock up before you're able to properly seat it. Next, glue the end and side cleats in place. The top is now screwed on through the slotted holes in the cleats.

We used a tung oil finish, because it combines the high gloss of a lacquer finish with the in-the-wood look of a penetrating oil. Flood on a heavy first coat, let dry, apply a second coat, let dry, and then rub with 000 steel wool. Now lightly work in one last coat of tung oil, let dry, and polish with 0000 steel wool to the final desired luster.

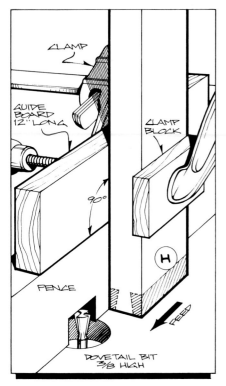

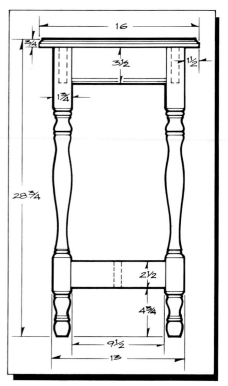

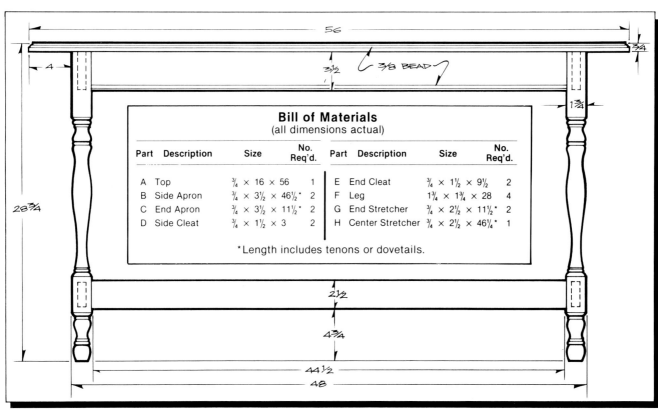

Bill of Materials
(all dimensions actual)

Part	Description	Size	No. Req'd
A	Top	¾ × 16 × 56	1
B	Side Apron	¾ × 3½ × 46½*	2
C	End Apron	¾ × 3½ × 11½*	2
D	Side Cleat	¾ × 1½ × 3	2
E	End Cleat	¾ × 1½ × 9½	2
F	Leg	1¾ × 1¾ × 28	4
G	End Stretcher	¾ × 2½ × 11½*	2
H	Center Stretcher	¾ × 2½ × 46¼*	1

*Length includes tenons or dovetails.

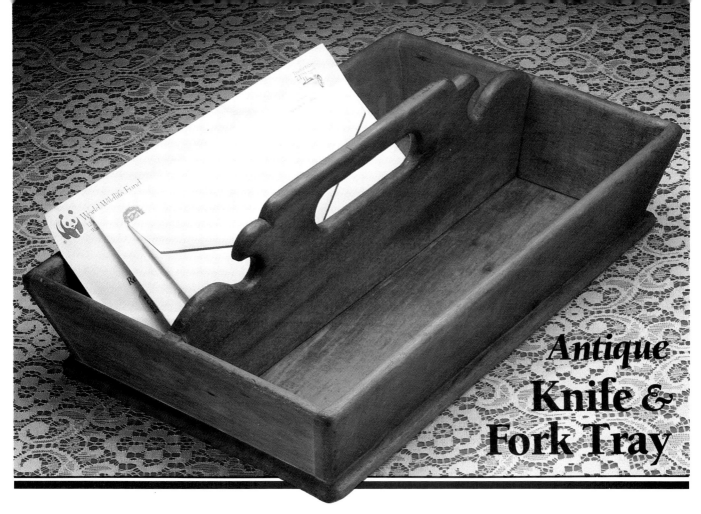

Antique Knife & Fork Tray

The needs of America's first settlers were few compared to ours today. Where a present-day kitchen would be considered naked without walls of cabinets, a Colonial kitchen might have contained only a water bench, table, and sideboard. Common flatware of copper, pewter, or latten (a tin covered iron; silver was found only in the homes of the wealthy) was typically held in a knife and fork tray.

Our tray is an excellent example of an antique knife and fork tray. Like many early pieces, it was crafted from native New England pine and held together with brads.

This is an ideal one-day project, since it can be completed in several hours. You'll need $\frac{3}{8}$ in. thick stock which you can either purchase pre-surfaced or resaw and plane yourself.

If you need to resaw, we recommend using the band saw for this operation. Band saw blades cut narrow kerfs, wasting less stock than table saw rip blades. Use as wide a blade as your band saw will accommodate. The workpiece should be flat, and the edge that rides on the band saw table should be jointed square.

It might seem logical to use a long, high fence as a guide for resawing, but most band saws do not cut absolutely true. Such a fence prevents you from making the small lateral adjustments to the board that are necessary to keep the blade cutting on the line as you feed the stock through. We recommend a simple pivot guide, as shown in Fig. 1, which is clamped in place at the proper distance from the blade to produce the desired thickness. The chamfered contact point of the guide enables you to shift the board slightly and thereby counteract any tendency of the blade to wander. A 1 in. by 8 in. by 28 in. long board, when resawed, will provide all the stock needed to make one tray, and light hand-planing will reduce the resawed material to the final $\frac{3}{8}$ in. thickness.

Once the various parts have been roughly laid out on the $\frac{3}{8}$ in. stock, cut the bottom (A) to length. Then rip and crosscut the stock for the two sides (B) and the two ends (C), allowing at least $\frac{1}{2}$ in. extra in both length and width for these parts.

The next step is to cut parts B and C to final length. Be careful, though, as some potentially confusing angles are involved. Because both the tray sides and ends tilt out at an 18-degree angle (shown in the elevation views), the butt joints at the corners where these parts meet will form a compound angle. The ends of both the side and end parts must be cut at this compound angle for the butt joint to be neat and clean. As illustrated in Fig. 2, by angling the table saw blade at $5\frac{1}{4}$ degrees, and setting the miter gauge at 17 degrees, you'll get the perfect compound angle cut required for the 18-degree tilt of the sides and ends. You must reverse the miter gauge setting to cut the opposite ends of all parts B and C.

Next, the side and end parts must be ripped to final width. Here, too, angles are involved. The top and bottom edges of parts B and C must be ripped at an 18-degree angle, equal to the desired tilt of these parts. Incline the table saw ripping blade at 18 degrees and make the ripping cuts as illustrated in Fig. 3. Note that Fig. 3 shows a saw blade that tilts to the left. If your saw tilts to the right make sure that the fence is on the right side of the blade. Be sure the edge bevel cuts are oriented properly in relation to the compound miter cuts made earlier.

To lay out the handle profile, position the handle stock under the full-size half-pattern with a sheet of carbon between the page and your stock. Then flip the page over and use the carbon to trace the other half of the pattern. You shouldn't have any trouble reading the profile through the page, but if you didn't press hard enough, use the carbon to first outline the profile on the page back.

Drill ⅝ in. and ¾ in. diameter holes as indicated on the full-size pattern to establish the inner radii, then cut out the handle's outer profile using a band saw, jigsaw, or by hand with a coping saw. The coping saw is also handy for completing the inner profile.

We used a router table with a ³⁄₁₆ in. radius bearing-guided round-over bit to round the bottom and the curved profile of the handle before assembly. As you will note in Fig. 4, the bit must be set at slightly less than the ³⁄₁₆ in. cutting height so the ball bearing guide will have a slight flat to bear off. Use a rasp and sandpaper to round the tops of the sides and ends after assembly.

To assemble the tray, first join the sides and ends with brads or finishing nails. Genuine cut nails or brads are available from The Tremont Nail Company, P.O. Box 111, Wareham, MA 02571. Next, add the bottom, and lastly the handle, using brads to secure these pieces also. Distress the tray with small nicks from a chisel, simulating years of wear.

Since our tray was an antique, we could only guess as to the finish and stain used, if any. For an authentic antique look, we recommend Minwax Early American or Special Walnut stain, followed by two applications of Minwax Antique Oil Finish. Rub out the finish with 0000 steel wool to the desired level of gloss.

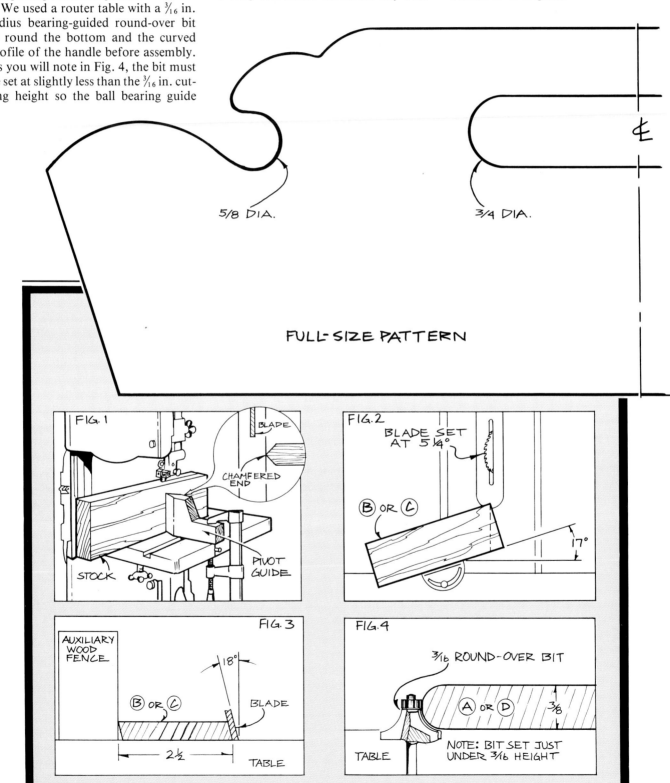

40 The Woodworker's Project Book

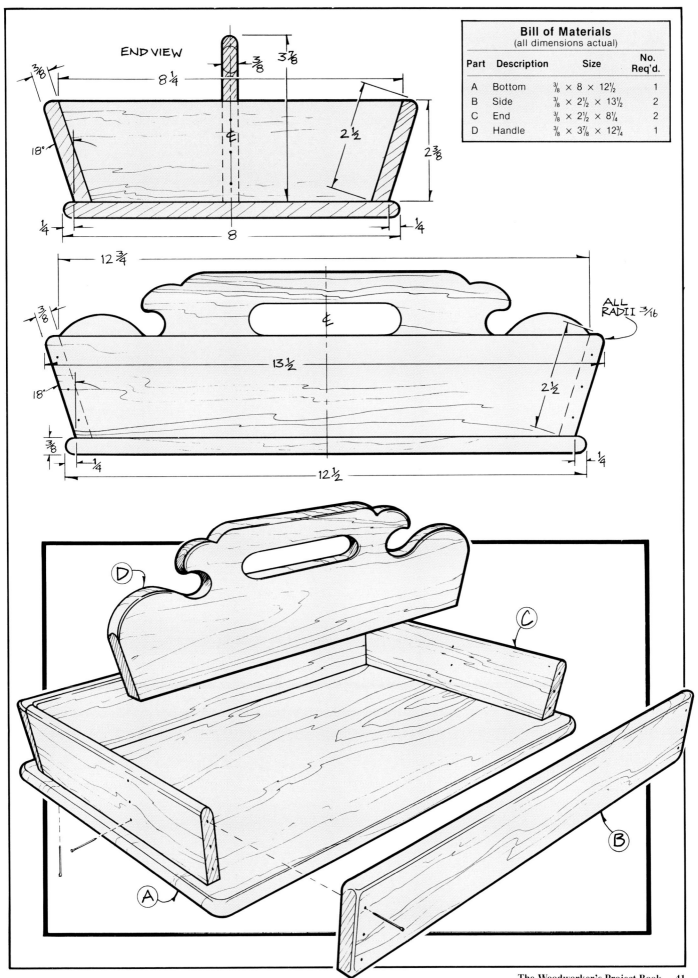

LOON • CARVING

by Rick and Ellen Bütz
Black and White photos by Ellen Bütz

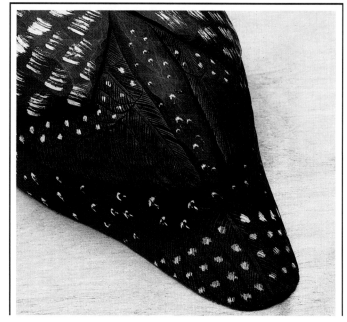

Photo 1: Transfer patterns to blocks.

Photo 2: Use the band saw to cut the side profile of the body.

Photo 3: Use the band saw to cut the top profile of the body.

Photo 4: Screw holding-block onto base of carving.

Photo 5: Rough out with a large gouge and mallet.

Photo 6: Outline the wings with a V-gouge.

The first loon I ever saw came drifting eerily out of the morning mist on an Adirondack lake. It saw me only a few feet away, called sharply, and dove without a ripple. For the first time, I understood why this bird has become a symbol of the wilderness. There was something wild and utterly primitive in the experience.

Loons are primitive birds, virtually unchanged from their fossil ancestors. Their heavy bones help give them the weight necessary to dive deep in pursuit of fish. They have been recorded as far as 200 feet below the surface.

Loons are large, powerful birds with few natural enemies. In recent years, however, man's activities have threatened them. The noise of motorboats drives these shy birds off their lakes. And the waves motorboats create can wash loons' low-lying nests off the lake shore. Even more serious, their food supply is threatened by the acid rain that kills fish in the high mountain lakes that have become their last refuge. This problem is more difficult to regulate than boating because it's the result of industrial smoke emissions originating hundreds of miles away.

Since that first sighting, loons have become one of my favorite subjects for wood carving. The pattern for this carving is scaled down to 12 in., about one-half life size. I've listed the sizes of the different tools that I use, but these are only guidelines. Other sizes will work; use what you have on hand.

To save wood and make carving easier, I make the loon from two pieces of wood. For the body, start with a 2¼ in. thick block of basswood or air-dried white pine about 4½ in. wide by 10½ in. long. For the head, use 1¾ in. thick wood 3 in. wide by 5 in. long. Lay the head pattern out on the wood with the grain running the length of the beak for maximum strength. Using the full-size patterns provided, trace both the side profile and the top profile of the loon body on the larger block, and the head profiles on the smaller block. Be sure the profiles line up (Photo 1).

Next, cut the pieces out on the band saw. When you cut out the body, cut the side profile first (Photo 2). That way, when you cut out the top profile, you will have a flat base to rest the carving on for safe, efficient sawing (Photo 3). Note that you'll need to redraw the top profile after you've cut the side profile on the band saw, or else tape the scrap piece with the top profile in place to serve as a guide. Only cut out the side profile of the head. The wood is thin enough so the rest of it can be whittled out with a knife.

To hold the wood while you're carving, screw a hardwood block into the base of the loon body. This lets you hold the wood in the vise on your bench (Photo 4).

For roughing out the body, I use a 35 millimeter no. 5 gouge. Just knock off all the corners. You can use a mallet for roughing out to carve away large chips quickly (Photo 5). You don't need fine control at this point. Don't make the neck area too thin; it will be shaped more completely after the head is glued on.

After you have the body rounded out, the next step is to define the wings. Use a 6 millimeter no. 12 V-gouge to incise along the outlines of the wings, which include the scapular, secondary and primary feathers (Photo 6). Then pare away below the cut with a 16 millimeter no. 2 gouge (Photo 7). This makes the wing appear raised above the body.

At this point, you're ready to start working on the head. Because the head is small enough to hold comfortably in my hand, I shape it mostly with a knife, rather than fastening it to the bench and using gouges. The knife I use is a small German carving knife. The shaped wooden handle is more comfortable to hold than a jackknife.

There are two basic knife cuts to use for fine control and safety. The first is the paring cut (Photo 8). Place your thumb on the block of wood and slowly close your hand drawing the knife through the wood. Keep your thumb low enough on the piece of wood so that the knife doesn't come into con-

Photo 7: Pare away below the V-gouge cut.

Photo 8: Shape the head using paring cut with knife.

Photo 9: Shape the head using levering cut with knife.

Photo 10: Detail the beak with a V-gouge.

Photo 11: Trim the beak with knife.

Photo 12: Cut a vertical line around the outline of the eye.

Photo 13: Shape the eyelid with a toothpick.

Photo 14: Blend body to head with the knife.

Photo 15: Sand the neck.

tact with it.

The levering cut is useful in places where it's awkward to position the knife for the paring cut (Photo 9). Place the back of the knife blade against your left thumb which serves as a fulcrum as you rotate the knife. This cut allows you to remove wood very precisely in tricky areas.

Remember, these are small controlled movements using only the muscles of your fingers and hand. Never pull or push the knife through the wood with a full arm movement.

After the head is shaped, incise a line with a V-gouge where the beak meets the face (Photo 10). Blend the V-gouge cuts smoothly into the face and beak with small knife cuts (Photo 11). Use the same techniques to incise a line separating the upper and lower portions of the beak.

Don't forget, a loon's beak is narrow and pointed like a spearpoint. It's not broad and rounded like a duck's spoon-shaped bill.

After you've finished shaping the head, the next step is to insert the eyes. Do this before you glue the head on the body, because it will be easier to position the eyes correctly. The bottom edge of the eye should not be any lower than the central line of the beak. Make sure the eyes are in the same position on each side of the head.

Use a small 4 millimeter no. 8 gouge to make a hole for the eyes — it gives a cleaner more precisely shaped hole than a drill. Start by cutting down vertically with the gouge all around the outline of the eye (Photo 12). Then, using the same gouge, scoop out the wood in the center. Keep the sides of the eyehole vertical.

The eyes I use for this size loon are 8 millimeter clear glass eyes that are made in West Germany. Because a loon's eyes are bright red, you should paint the back of the eye with red acrylic paint before inserting.

To hold the eye in place, I use a two-part epoxy putty. It fastens the eye securely into the head, and can also be shaped around the eye to form a very realistic eyelid. Be sure to mix the putty well or it won't harden properly. Place a small ball of putty in the eyehole, then gently press the eye down into it. Shape the excess that squeezes out around the eye with a toothpick to form the eyelid (Photo 13).

Now it's time to fasten the head to the body. The gluing surfaces must be

Photo 16: Use a chisel-tip burn-in pen to outline the feathers.

Photo 17: Burn in the feather barbs.

Photo 18: Burn in the beak details.

Photo 19: Loon — carved, detailed, lacquered.

Photo 20: The loon with the large areas of black and white painted on.

Photo 21: Dry-brush a feathered edge between the black and white areas.

Photo 22: Painting black stripes on the neck.

Photo 23: Painting on the white spots using dry-brush technique.

Photo 24: Finished loon.

smooth and flat to insure a solid invisible joint. Use a disc sander to get the most level surface. I use a quick-setting epoxy glue for this job. It doesn't require clamping; you can hold the two pieces in place for the five or so minutes it takes the glue to set. In fact, tests have shown that quick-setting epoxy glue bonds more strongly if not clamped. Also, the epoxy forms such a strong bond that you don't need to dowel the pieces together as you would with a water-base carpenter's glue. Turn the head a bit to one side or the other for a more lifelike look.

When the glue has hardened, shape the neck with a knife to blend the head and body pieces together (Photo 14). Then smooth the head and neck with sandpaper (Photo 15). Start with 150-grit garnet or aluminum oxide sandpaper and finish with 220-grit. This leaves a fine smooth surface. Pay special attention to the area where the head and body join. This joint must be sanded perfectly smooth or the glue line will show when the carving is painted. Usually, I prefer to leave the tool marks in the wood and seldom use sandpaper. However, the feathers of the head and neck are so fine on the loon that sanding works well.

Once the body is completed you may want to add some feather detail before painting. One way to do this is with a woodburning pen. I only burn in the larger feathers on the tail and the primaries and secondaries on the wings. On a loon, the rest of the feathers are so fine that heavy texturing is unnecessary.

Outline the feathers with a chisel-tip woodburning pen (Photo 16). This creates the step-down effect between overlapping feathers. Make sure your feathers overlap the right way. The higher ones go over the lower ones like the shingles on a roof. Use the same techniques to outline the tail feathers. If you don't have a woodburning pen, you can detail these feathers with a V-gouge. The technique is the same as the one used for defining the wing shape, but remember to make these cuts very shallow.

To make the barbs on the feathers, use a skew tip woodburning pen sharpened to make a very fine line (Photo 17). This is delicate work, so be patient and take your time.

Use the same tip to burn in the nostril and the line between the upper and lower portions of the beak (Photo 18).

(continued on next page)

After the carving is burned in, it should be sealed to protect the wood from changes in humidity and to provide a good surface for painting. I use Deft, either the aerosol spray or brushing lacquer (Photo 19).

The next step is painting. A loon's summer coloring is basically simple, just black and white. Look at some good photographs of loons to help you with correct placement of the colors. Acrylic paint is a good choice for this project as it dries quickly and covers well. On the loon, I use the paint thick, just as it comes out of the tube, for better covering power. If your paint is too thick to brush on smoothly, thin it very slightly with water.

I begin by painting in the large areas of black and white before adding the details. First, paint the front and sides of the neck and chest white. It may take two coats for complete coverage. Then paint the rest of the bird, including the beak, black (Photo 20). Use two coats if you need to. Don't worry about any paint that gets on the eyes. You can scrape it off easily with a knife point after the paint is dry.

Wait until the paint dries, then very lightly take some black paint on a small flat-bristle brush, like a Grumbacher no. 3 bristle, and gently dry-brush a little black paint into the areas of white in the direction of the feather growth (Photo 21). This creates a soft feathered edge between the two colors.

Dry-brushing is a painting technique where you have just enough paint on your brush to barely hold the bristles together. This way, when you paint a stroke with your brush, each individual bristle leaves a fine line rather than a glob of paint. Experiment on a piece of paper until you get a feel for just how much paint to leave on your brush.

The next step is painting the black stripes on the loon's neck. Use a no. 1 pointed sable or synthetic sable brush for the finest stripes, and a no. 4 for the broader stripes farther down the neck. Paint a very fine line, but don't make it perfectly straight; it will look more like feathers if it's broken up a bit (Photo 22). This necklace, as it's called, varies from loon to loon and is as distinctive as a fingerprint. It is a great aid to naturalists studying loons to be able to tell the individuals apart.

The final step is painting the white spots on the back and sides. The smaller spots on the side are done with a no. 3 pointed sable brush. The larger ones on the wings are done with the no. 3 flat-bristle brush used earlier. Use the dry-brush technique to make the spots soft and feathery (Photo 23).

After all the paint is completely dry, use a clean horsehair shoe brush to lightly buff the back, the burned-in areas of the wings and tail, and the beak. This adds just a little bit of shine to the black paint to make the tool marks more visible and duplicate the sheen of real feathers.

Making a carving of a bird or other wild animal is always exciting because it allows you to bring a bit of nature indoors. Remember, take your time and enjoy yourself. I think you'll find this a very rewarding project.

Rick and Ellen Bütz live in Blue Mountain Lake, New York. They are professional woodcarvers and authors of the book Woodcarving with Rick Bütz *(Madrigal Publishing). Recently, they completed a PBS television series, also entitled* Woodcarving with Rick Bütz.

Woodcarving Tools

Garrett Wade
161 Avenue of the Americas
New York, NY 10013

Wood Carvers Supply, Inc.
P.O. Box 8928
Norfolk, VA 23503

Woodcraft Supply
P.O. Box 1686
Parkersburg, WV 26102

Glass Eyes

P. C. English
P.O. Box 380, Dept. WJ
Thornburg, VA 22565

Christian Hummul Company
404 Brookletts Avenue
P.O. Box 1849, Dept. WJ
Easton, MD 21601

FULL-SIZE PATTERN

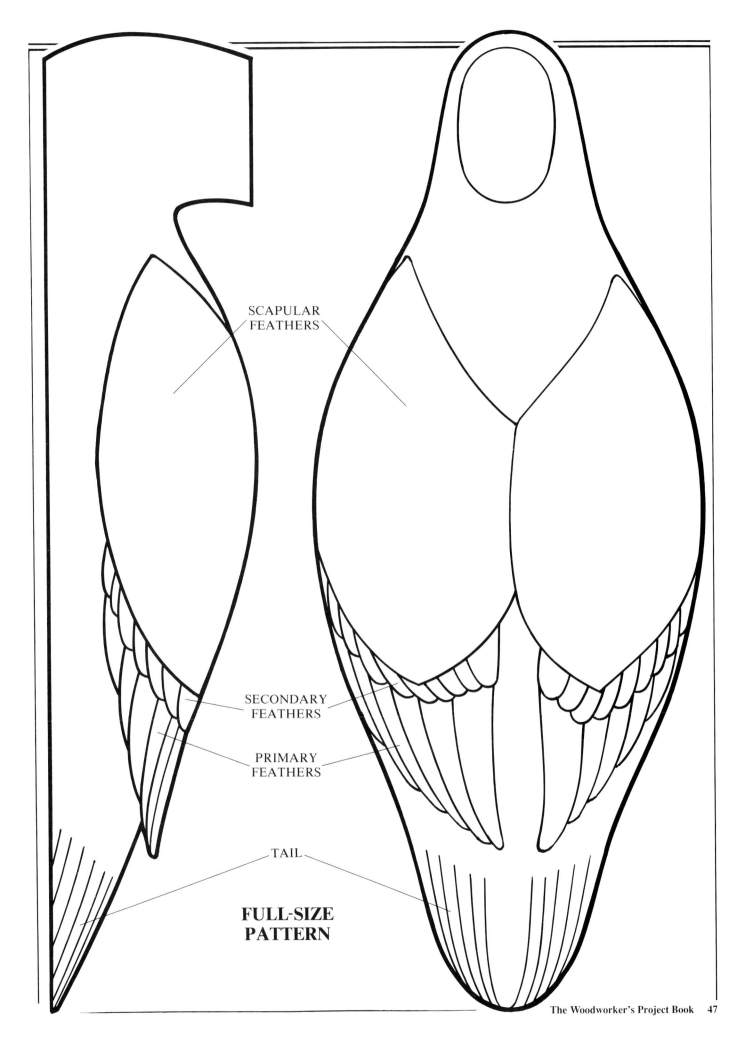

Old-time Pipe Box

This pipe box offers plenty of practice with dovetail joints. The pine original, an antique, has dovetails at all four corners of the small drawer. You may choose to cut just two at the drawer front, a common arrangement, or to cut all four just for fun.

First cut the base (A) and box parts to size, and transfer the profiles shown in the grid patterns to the back (B), front (C), and sides (E). Cut with a coping saw or a jigsaw, being careful to stay on the waste side of the line. Then round over the base board with a 3/8 in. round-over router bit or a block plane.

Because you'll want to make the drawer to the size of the actual opening, next sand and assemble the box with finishing nails as shown.

Then cut the drawer parts to size, and establish the 1/4 in. deep rabbets on the inside of the drawer front (F). They are 1/4 in. wide on the bottom, but 3/16 in. wide on the sides and top. Orient the parts and label them.

Next, scribe the depth of cut for both tails and pins with a marking gauge. (The tails are the portion that resemble a dove's tail. The pins are the matching beveled shoulders.) Set the gauge a hair over 1/4 in., the thickness of the drawer sides (G). Scribe lines on the inside of the drawer front (from the edge of the rabbet), on both ends of the drawer back (H), and on both ends of each side. When marking the sides and back run the lines on all four surfaces.

After scribing the depth cuts, mark the tails on the sides using a bevel gauge or a template. Run the lines down to the depth marks and also across the end grain. Use a scratch awl or sharp knife. Mark the waste sections with an "X".

Use a dovetail saw to cut the tails to depth. Stay on the waste side of the line. After cutting the profile, make several relief cuts in the waste material.

With a coping saw or chisel, cut along the depth line and remove the waste. Clean the cuts with the chisel. Next, trace the tail profiles onto the end grain of the drawer front and the back. Hold the piece you're marking in a vise. Use a try square to run lines from the end grain to the depth marks.

Also use the dovetail saw and chisel to cut out the pins. For the half-blind dovetail you won't be able to cut all the way through with a saw. However, you can make a triangular cut from the inside. Hollow out the remainder of the cut with a chisel, cutting first across the grain and then gently lifting out the chips while cutting into the end grain.

After cutting the dovetails, make the 1/8 in. deep by 1/4 in. wide rabbet in the drawer sides to accept the bottom (I). Also cut the 3/16 in. bead on the face of the drawer.

Before final assembly, dry fit the drawer, knocking the parts together lightly. You may need some chisel work to get the dovetails to fit correctly. They should go together snugly. If you have to bang on them, they'll split after gluing.

Next, cut a 3/8 in. diameter hole in the back as shown and drill a shank hole in the drawer front for knob (J).

Sand well and stain with two coats of Minwax Cherry Wood Finish. If desired, nick the box with the corner of a chisel for a distressed look. Finish with shellac.

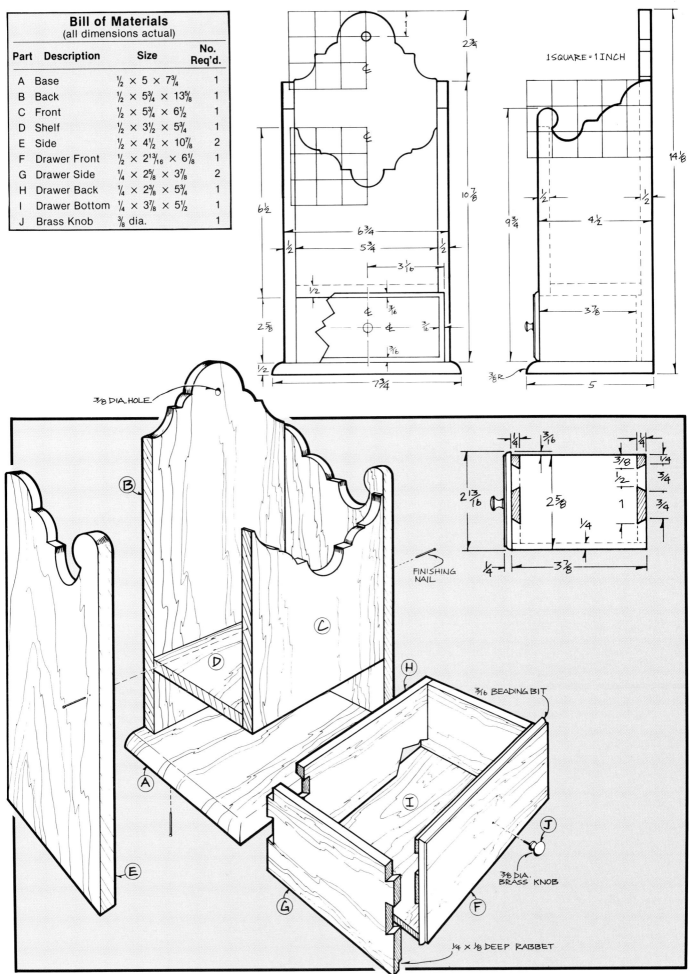

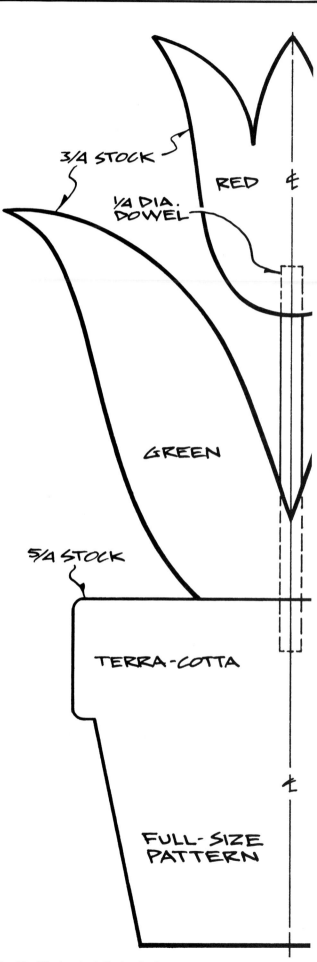

Dutch Tulip
Folk-Art Silhouette

This cheery silhouette will add a touch of charm to your kitchen window or sun porch. It's easy to make and could even be a family project. Mom or Dad can cut the pieces and lay the newspaper on the kitchen table. Everybody else can pitch in with the sandpaper and paint.

The flowerpot base is made from 5/4 pine. The leaf and flower are 3/4 in. pine. A 1/4 in. diameter by 4 in. long dowel connects the three parts and serves as a stem.

Use our full-size pattern to transfer the profile onto the three wood blocks. Then drill the 1/4 in. hole through the leaf section and 1/2 in. deep into the flower and base.

You can use a band saw or jigsaw to cut out the parts. Clean up the cuts with a file, and sand well with 150- and 220-grit sandpaper before finishing.

Apply a base coat of primer or flat white paint before the colors. The primer sanded with 220-grit paper provides an excellent surface for the enamel colors.

We used gloss enamel paint for all colors: red for the flower and green for the leaves. To give the pot its terra-cotta shade we mixed four parts each of red and yellow, with three parts of white, and one part of blue. If you need two coats of color, sand the first coat well.

After the paint dries, apply glue to the ends and middle of the dowel and assemble the sections.

Four-Drawer Lamp

Although as a chest of drawers this lamp won't replace your wardrobe, it does provide a small place for tiny things, and there's plenty of light to find them by.

Start by preparing enough stock for the various parts. You'll need $\frac{3}{4}$ in. pine for the base (A) and drawer fronts (F), and $\frac{1}{2}$ in. pine for the top (B), shelves (C), sides (D) and back (E). You'll also need $\frac{3}{8}$ in. pine for the drawer sides (G) and the drawer backs (H), and $\frac{1}{4}$ in. plywood for the drawer bottoms (I). If the thin pine stock isn't available locally, resaw it from $\frac{3}{4}$ in. boards. The pieces are narrow enough to resaw on the table saw with a high rip fence. You'll need to run them through twice since most table saws won't make the widest cut in one pass. Keep the same face against the rip fence for both passes. Resaw the stock $\frac{1}{32}$ in. oversize and plane it down to the final size.

After getting the stock ready, cut the case parts to size. Put the drawer stock aside because you'll make the drawers to fit after finishing the case.

When cutting the parts, rip the stock to size first, then crosscut the parts. For the ripping operation, leave the parts slightly oversize and joint them smooth. For crosscutting, we recommend a saw blade with fine teeth for a smooth cut, as well as a stopblock for uniformity. Hold a scrap block against the trailing edge of the workpiece as you run it past the sawblade to prevent chipping. The dado grooves in the sides extend to the face, so any chipping along the edge will show in the finished piece.

Cut the shelves just a hair over the dimension shown. You'll sand the edges with a sanding block before assembly, and the soft pine sands easily.

Next, cut the $\frac{1}{8}$ in. deep by $\frac{1}{2}$ in. wide open dado joints in the sides. We used a $\frac{1}{2}$ in. straight bit in the router. You can also use a dado blade or make multiple passes with a good crosscut blade in a table or radial-arm saw. However you make the cuts, keep them uniform by using a stopblock or fence, and back up the pieces with scrap to prevent chip out. Pine splinters very easily. Accuracy is especially important here because the cuts establish the drawer openings. Make sure you have uniform dimensions between the grooves, that the depth is precise, and that the width of the cut matches the stock thickness. It's a good idea to make some test cuts in scrap to get the setup right.

Although there are five shelves, you only need to adjust your fence or stopblock for three cuts. Once for each end, once for the two shelves above and below the ends, and once for the middle shelf. Success with this procedure depends on accurate stock size. If one side is slightly longer than the other, the shelves will slant. Be sure the lengths are equal before you cut.

Cut the $\frac{1}{4}$ in. deep by $\frac{1}{2}$ in. wide rabbet for the back next. You'll have to change your setup to get the wider and deeper cut. You can adjust the fence on a router table, but with a table or radial-arm saw, you'll have to change from crosscutting to ripping.

To cut the profile on the base and top we used a $\frac{3}{8}$ in. beading bit in the router. You can use a hand-held router with a bearing-guided bit, or a router table setup without the bearing. When cutting, back the workpiece up with scrap to prevent chip out. Also start with an end-grain side and rotate the pieces for successive cuts. The last cut will be on a long-grain side and shouldn't splinter.

Next, cut the $\frac{1}{4}$ in. deep by $\frac{1}{4}$ in.
(continued on next page)

wide grooves for the electrical cord. We used the router table with the ¼ in. straight bit. Make sure your parts are labeled and that you cut the proper faces of the back and top shelf. For the base, cut a ¼ in. by ¼ in. slot on the bottom (see section view) and stop the cut 1 in. in from the edge. Drill a ¼ in. hole in the top of the base to meet the slot.

Finish the case by cutting the 1½ in. diameter hole in the top shelf and the ⅜ in. diameter hole in the top for the fixture. The larger hole in the top shelf won't show and provides working room when assembling the lamp.

Next, sand the parts and assemble the case. First dry-fit the parts to check for squareness. Then apply white or yellow glue to the joints and clamp. Leave the base off and plan to screw it onto the finished lamp. That makes it easier to wire the lamp.

The drawers are made with a simple rabbet lip, as shown in the drawer detail. Make sure to measure the actual drawer openings before cutting the pieces.

Use the router and router table to cut the ¼ in. wide by 3/16 in. deep grooves for the drawer bottoms (I). The groove is ⅛ in. from the bottom edge of the sides (G) and back (H), but ¼ in. from the bottom edge of the drawer front (F).

Use the router again to establish the rabbet lip on the inside of the drawer front to match the sides. The ⅜ in. depth remains the same for all four sides, but the width of cut is ½ in. on the sides, ⅛ in. on the top and bottom.

Cut the rabbet lips individually for each drawer and, to get a good fit, make the ½ in. cut on the sides last. Start a little long and gradually adjust the setup until the drawer fits right. To test fit, make a cut and hold the front between two sides as you try it in the opening.

Cut the drawer front profile with the ¼ in. beading bit in the router. Again, back up the cuts with a scrap block and make an end-grain cut first. The ¼ in. plywood bottoms are cut a hair under the actual groove-to-groove dimension. That prevents the bottoms from interfering with a tight fit of the drawer cases.

The drawers are assembled with ½ in. brads and glue. The bottom panel floats in the grooves, so don't use glue on it.

Sand the drawers and the case thoroughly with 150- and 220-grit paper. We used Cook and Dunn Early American stain and shellac to finish the lamp. Hand rubbing the shellac with pumice and rottenstone provided a soft luster.

To attach the knobs (R), drill shank holes for the screws, measuring carefully to find the drawer centers.

We've arranged for all the lamp hardware shown, except for the shade and knobs, to be provided by a single supplier. The hardware kit, which can be ordered from Craftsman Wood Service, 1735 West Cortland Court, Addison, IL 60101, includes an 8 ft. long brown cord (P), a push-through switch socket (O), a 12 in. harp (J), a 1 in. long threaded nipple (K), a ¾ in. long threaded nipple (L), a 2⅝ in. tall turned brass spindle (N), a 2 in. tall turned brass finial (Q), and one brass nut (M). Be sure to specify part no. WWJ-1. You should be able to find the shade and brass knobs locally.

The hardware kit is designed for use with a 14 in. tall shade, which we purchased at a local department store. The 14 in. shade is a common size and is available in different fabrics and colors. We selected burlap since it seemed to complement the natural wood well. If you do decide to use a different size shade, the size of the harp will probably have to be altered to match.

The various hardware components are assembled as shown in the exploded view. Thread the lamp wire up through the lamp, starting at the bottom. Several insulated staples will secure the lamp wire in the ¼ in. groove in the lamp back. Finally, screw the lamp base in place.

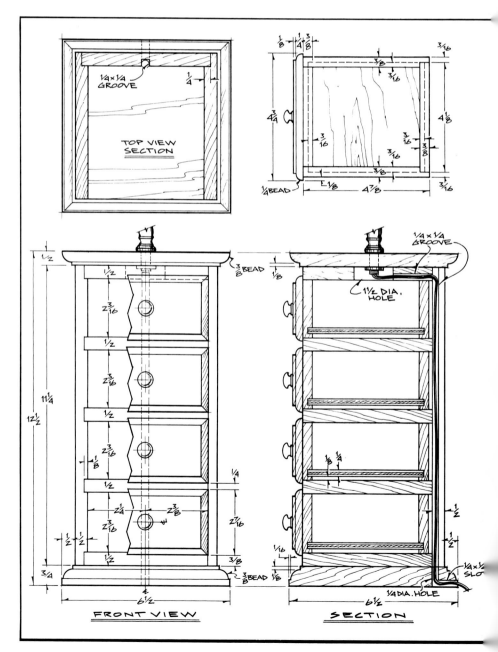

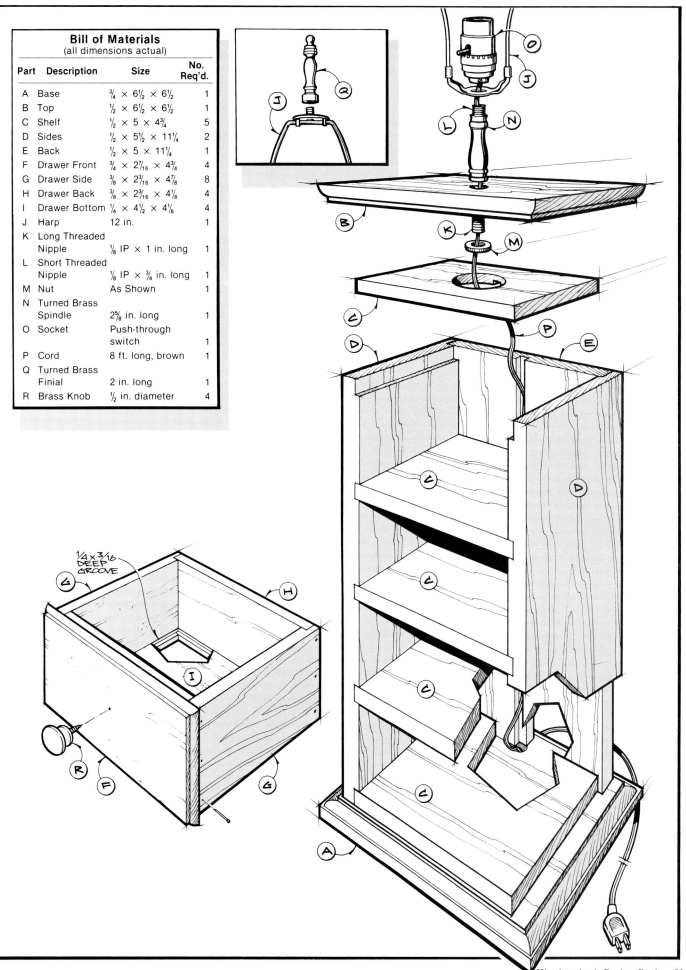

Bill of Materials
(all dimensions actual)

Part	Description	Size	No. Req'd.
A	Base	$3/4 \times 6{1/2} \times 6{1/2}$	1
B	Top	$1/2 \times 6{1/2} \times 6{1/2}$	1
C	Shelf	$1/2 \times 5 \times 4{3/4}$	5
D	Sides	$1/2 \times 5{1/2} \times 11{1/4}$	2
E	Back	$1/2 \times 5 \times 11{1/4}$	1
F	Drawer Front	$3/4 \times 2{7/16} \times 4{3/4}$	4
G	Drawer Side	$3/8 \times 2{3/16} \times 4{7/8}$	8
H	Drawer Back	$3/8 \times 2{3/16} \times 4{1/8}$	4
I	Drawer Bottom	$1/4 \times 4{1/2} \times 4{1/8}$	4
J	Harp	12 in.	1
K	Long Threaded Nipple	$1/8$ IP $\times$ 1 in. long	1
L	Short Threaded Nipple	$1/8$ IP $\times 3/4$ in. long	1
M	Nut	As Shown	1
N	Turned Brass Spindle	$2{5/8}$ in. long	1
O	Socket	Push-through switch	1
P	Cord	8 ft. long, brown	1
Q	Turned Brass Finial	2 in. long	1
R	Brass Knob	$1/2$ in. diameter	4

Salt Box

If there were a "top ten" listing of all-time favorite Early American projects, this small salt box would certainly have to be included. We've seen many different salt box designs, but this particular style seems to have just the right proportions, highlighted by the classic decorative back profile.

Perhaps best of all, this project doesn't require a shop full of fancy equipment or years of woodworking experience. Just about anyone who takes the time to carefully measure and cut, and follow the grid for the back profile, will end up with an attractive finished product.

As you'll note from the elevation views, all parts are made from ½ in. thick stock. You may have to hand-plane ¾ in. thick stock to reduce it to the ½ in. dimension, which entails some work, but we don't recommend that you substitute ¾ in. material for the ½ in. designated thickness. If you build the piece from ¾ in. stock, you'll find that the box looks "clunky."

After you've planed enough material for all parts, cut the bottom (A) to size, and rough-cut stock for the back (B). Then transfer a 1 in. grid pattern to the upper half of the back, draw in the curved profile, and cut out. You could use the jigsaw or band saw to cut this profile, although a much better choice would be a scroll saw or a coping saw. With a little practice, most woodworkers discover that the coping saw is an extremely effective tool for this type of work.

Next, cut the sides (C). Once again, the coping saw will come in handy for cutting the upper end of the sides to the designated profile. Drill a ¼ in. diameter hole as shown for the pivot dowels that will be used to mount the lid.

After cutting the front (D) to its 7 in. length and about 5 in. width, assemble the box using finishing nails or brads, and glue where the back and sides meet. Hand-plane the top edge of the front to match the slope of the sides.

The lid (E) is cut to length and width and then notched to fit between the sides. While you could use the table saw to establish the 22-degree angle on the bottom edge of the lid, a few passes with the hand plane will get the job done just as quickly without having to spend time setting up the table saw. As an exercise in using hand tools, this project could be a good way to rediscover the joy of doing things entirely by hand, without the ear-splitting whine of the table saw or router.

Lastly, you'll need to drill the holes in the lid to accept the pivot dowels. Position the lid and, using the previously drilled holes in the sides as a guide, drill the ½ in. deep holes to accept the dowels. If you've got a brace and a set of auger bits, you might use them to keep this project an entirely handmade piece. To assemble, place a little glue on one-half of each dowel, then insert the ends without any glue through the sides and into the lid. Take care not to get any glue on the dowel ends that fit into the lid. Also, note that those ends may have to be filed or sanded slightly; if they bind tightly, the lid won't pivot freely.

Minwax Special Walnut stain gives pine a nice aged look. One coat of Minwax Antique Oil finish completes the project.

Bill of Materials
(all dimensions actual)

Part	Description	Size	No. Req'd.
A	Bottom	½ × 4¾ × 7½	1
B	Back	½ × 7 × 12	1
C	Side	½ × 3½ × 6¾	2
D	Front	½ × 4⅞ × 7	1
E	Lid	½ × 4¼ × 7½	1

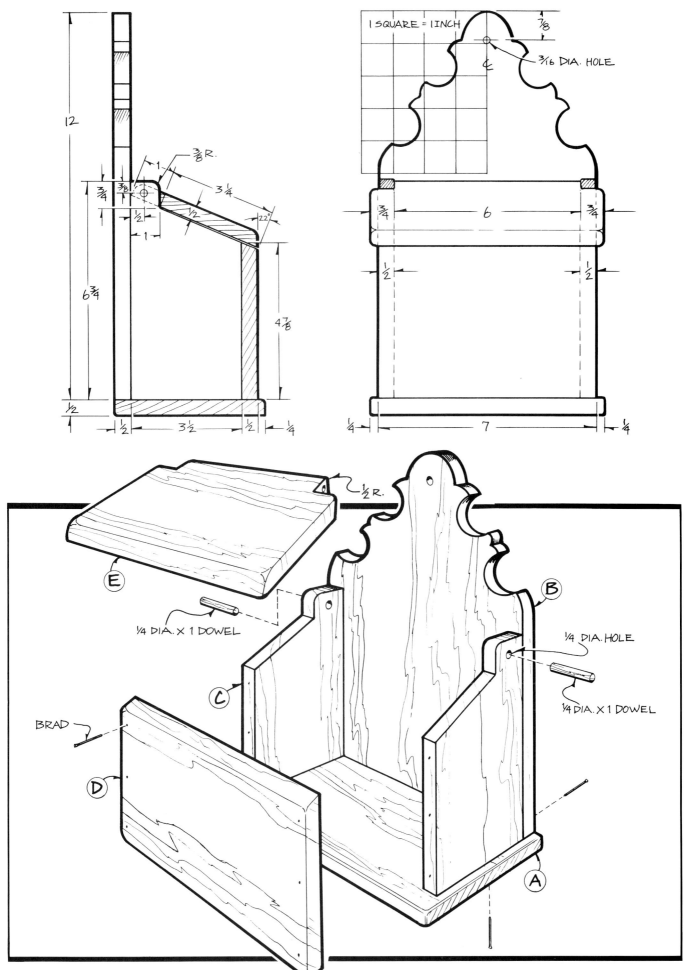

Bud Vase

We like this bud vase for its elegant profile and for the simple details that make it a good practice piece for novice woodturners.

We made it from cherry, a wood with a rich color and interesting grain. Walnut, rosewood, or a figured birch would also work well.

Before starting, transfer the profile of the full-size pattern onto something stiff to use as a template: 1/4 in. plywood or hardboard. You only need one edge to use as a guide, so cut the profile in the top but leave the bottom edge straight. Mark the dimensions of beads, coves and V-grooves, as well as the maximum and minimum diameters (Fig. 1).

Next, secure a turning block, find the centers, and mount it in the lathe. The blank should be at least 1 in. longer than the finished piece.

We've arranged the subsequent operations in a step-by-step format to help beginners through the cutting steps.

Fig. 1: Rough out the cylinder leaving a 2 in. diameter at the headstock as shown, but turning it down to just over 1½ in. diameter starting at a point about an inch in. Make the transition from 2 in. to 1½ in. a gradual one. With the workpiece stopped, use calipers to check the diameters.

On the roughed-out cylinder mark the locations shown: the shoulders of the coves, the sides of the bead, the bottom of the long hollow, and the top of the vase. Hold the full-size template over the revolving turning and transfer the lines with a pencil.

After transferring the lines, place the template directly behind the turning in such a way that you can refer to the top profile as you work.

Fig. 2: Use a parting tool to cut to depth at the marks, following the diameters shown on the template. Cut on each side of the bead lines, leaving the raised area. For the cove shoulders, cut the higher one first, then the lower one. Note that your parting tool may be wider than the shoulder. On the higher shoulder it won't matter because you'll cut down on both sides. But on the other shoulder be careful to cut on the left side so the tool doesn't eat into the vase-shaped taper. When cutting with the parting tool, remember it doesn't make a finish cut, so you'll need extra wood for cleaning up later. Also cut the nub on the right to about ½ in. diameter with the parting tool.

Fig. 3: Round over the bead with the skew chisel. For this operation you should increase the lathe speed to about 2,000 rpm. Cut slowly, working first at the very corner and taking light shearing cuts until you're cutting all around the bead. Remember to use the short corner of the skew and to position the tool so the bevel rubs before you begin the cut.

Fig. 4: Working with the medium or large gouge, rough out the vase shape on both sides of the bead. Use a shearing cut and always cut downhill, from larger to smaller diameter. After roughing out the vase shape, smooth it with light shearing cuts of the skew chisel, again always cutting downhill. Then use the skew to slice away the wood for the bevel at the top of the vase. Use the calipers often to check dimensions.

Fig. 5: Use the small gouge to cut the coves. First hog out the waste material and then use a shearing cut to form the profile. Remember to start the cut with the gouge on its side and roll it into the cut, ending with the gouge on its back. Work from both sides of each cove. Set the calipers for the final depth of each cove, and work until you reach that depth.

Cut the V-grooves with the short corner of the skew chisel, again working from both sides.

Next, remove the tool rest and, if needed, sand the turning with 150- and 220-grit paper.

Finally, cut the finished vase out of the turning block with the parting tool. Cut straight in at both ends, leaving only about 1/4 in. of wood. Remove the workpiece from the lathe and carefully cut off the remaining nubs with a handsaw. Then sand the base and top smooth. Use a brad-point drill bit for the 1/4 in. diameter opening in the top. Countersink the hole and finish the bud vase with lacquer or penetrating oil.

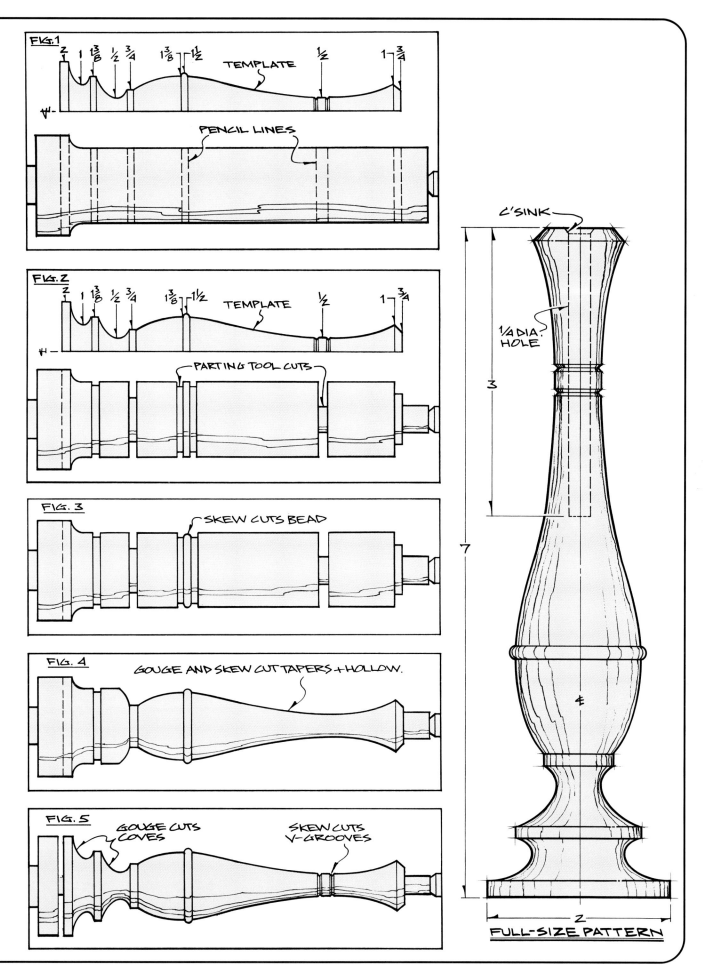

Early American Dry Sink

With its country flavor and simple charm, this dry sink can be a kitchen's friendly catchall, or a parlor's plant-filled island.

The design is simple. It's made in two sections, much the same as a hutch. Here, however, the stacking upper unit is replaced by three drawers and the open sink. The base is merely a box with facing boards screwed to the front, and a door that fits flush with the face.

When making the dry sink, it's a good idea to construct the base and upper sections first, then make the drawers and door to fit the actual openings. However, you'll use the same setup to cut mortise-and-tenon joints in the base and in the door frame, so plan your operations carefully to save time.

Start by edge-joining boards for the wide panels. Unless you have access to very wide stock, you'll need to join boards for the sides (A) and bottom (B) of the base, the adjustable shelf (F) and the raised panel of the door (K). You'll also need to join boards for the bottom (N), sides (O), and back (R) of the upper section.

The grain runs vertically for the base sides, but horizontally for the upper-section sides. When edge-gluing, remember to make the pieces a little bigger than needed, and cut them to exact size after the glue dries. We don't use dowels because they weaken a long grain-to-long grain joint.

Next, cut the rest of the parts to size. Leave a little extra wood on the upper-section parts that fit together for the 10-degree angle: the bottom (N), the side (O) and the front (S). That allows you some leeway in case the angles, which are cut later, aren't perfect.

Once the glued-up panels are dry and cut to size, plough the dadoes and rabbets in the base sides (A), upper-section sides (O), and the shelves (P). You can use a dado head in your table or radial-arm saw, or rabbeting bits in a router.

The sides (A) have a ⅜ in. deep by ¼ in. wide rabbet for the back (G), a ⅜ in. deep by ¾ in. wide by 2 in. long notch for the stretcher (H), and a ¼ in. deep by ¾ in. wide dado for the bottom (B).

The upper-section sides (O) have ⅜ in. deep by ⅜ in. wide grooves for the bottom (N), and ¼ in. deep by ¾ in. wide grooves for the shelves (P) and back (R).

The upper-section shelves (P) have ¼ in. deep by ¾ in. wide dadoes for the drawer dividers (Q). The bottom (N) has a ⅜ in. long by ⅜ in. thick tongue that fits into the sides (O).

Cutting the various grooves, dadoes and rabbets goes fairly fast if you first make all the ¼ in. wide cuts, then the ⅜ in. cuts, then the ¾ in. cuts. However, you'll need to cut the notches for the stretcher (H) with a chisel.

Next, lay out and drill the ⅜ in. diameter holes for the ¾ in. long adjustable-shelf pegs. Make the holes ⅜ in. deep. Then transfer the grid pat-

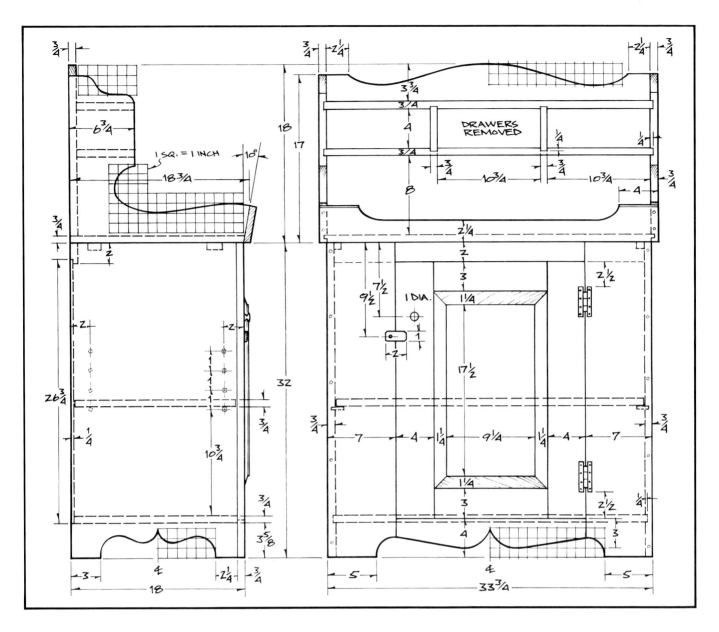

terns to the stock and cut the profiles with a scroll saw or jigsaw. Clean up the cuts with a drum sander or wood files. It's best to hold off cleaning the profiles in the stiles (C) and bottom rail (E) until after you put the face frame subassembly together.

Now make the 10-degree slope on the front end of the upper-section sides (O) with a table saw. Also use the table saw to cut the 10-degree slopes on parts N and S. We suggest dry-fitting upper section parts N and S after cutting the slopes in the sides (O). That way, if your first angle is slightly off, you can cut the others to match. We also suggest planing the angle on the top edge of the front (S) after assembly, a procedure that insures it will match the curve of the side.

Next, cut the mortises in the base stiles (C) and the tenons in the rails (D and E).

Note that the mortises and tenons for the base and the door frame are pretty much the same, so you may want to cut the base joints, assemble the carcase, and use the same setup for the door joints. That way you'll make the door to the actual opening and not waste setup time.

To cut the ¼ in. wide by ⅜ in. deep grooves in the door rails and stiles to accept the raised panel, use a ¼ in. straight cutter in the router table. Set the fence ¼ in. from the cutter. However, don't make the ⅜ in. deep cut in one pass. You'll get a smoother cut, with less strain on the motor, if it's done in three passes, with each pass removing ⅛ in. of material.

The ¼ in. wide mortises in the door and base stiles are cut a shade over 1⅛ in. deep. You can use the router setup to establish the dimensions, and then deepen them with a chisel. Or you can cut them entirely with a mallet and chisel. Use a marking gauge when laying them out to insure uniformity.

When laying out and cutting the mortises and tenons, remember to decide beforehand which sides are faces and mark them as such. Then keep the same side out for all the marking and cutting. If you flip the pieces face-for-face, any small variations in wood thickness will be doubled and show up as sloppy joints. Here the mortise and both shoulders are ¼ in. Don't let the seeming symmetry trick you into turning the pieces over and ruining a joint.

For the tenons, use the table saw with a tenoning jig. First establish the shoulders, cutting carefully with a stopblock on the miter gauge as shown in Fig. 1. Remove the rest of the shoulder with a tenoning jig, as shown in Fig. 2. Re-
(continued on next page)

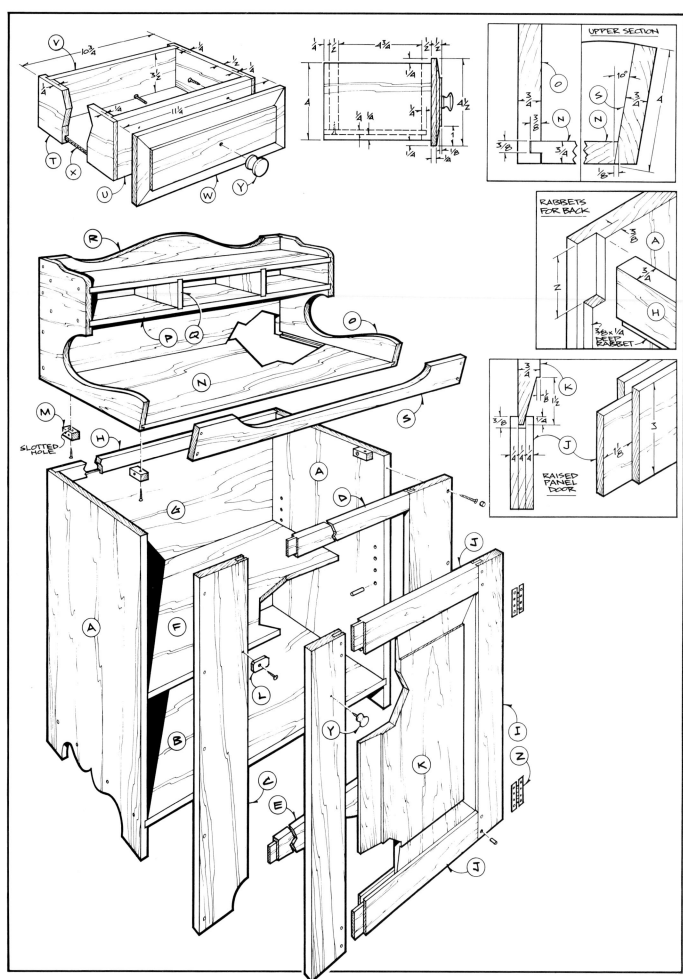

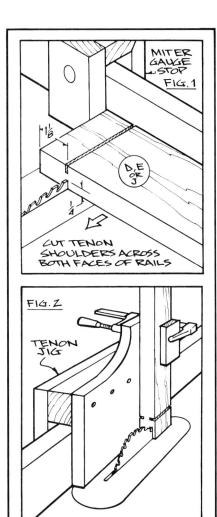

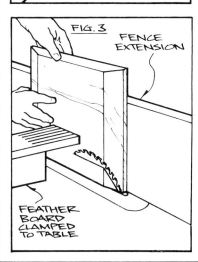

Bill of Materials
(all dimensions actual)

Part	Description	Size	No. Req'd.
Base			
A	Side	3/4 × 17 1/4 × 32	2
B	Bottom	3/4 × 17 × 32 3/4	1
C	Stile	3/4 × 7 × 32	2
D	Top Rail	3/4 × 2 × 22*	1
E	Bottom Rail	3/4 × 3 × 22*	1
F	Adjustable Shelf	3/4 × 16 3/4 × 32	1
G	Back	1/4 × 26 3/4 × 33	1
H	Stretcher	3/4 × 2 × 33	1
I	Door Stile	3/4 × 4 × 26	2
J	Door Rail	3/4 × 3 × 14*	2
K	Door Panel	3/4 × 12 1/4 × 20 1/2	1
L	Door Latch	1/4 × 1 × 2	1
M	Cleat	3/4 × 1 × 2	4
Upper Section			
N	Bottom	3/4 × 18 1/8 × 34 1/2	1
O	Side	3/4 × 17 × 18 3/4	2
P	Shelf	3/4 × 6 × 34 1/4	2
Q	Divider	3/4 × 6 × 4 1/2	2
R	Back	3/4 × 17 1/4 × 34 1/4	1
S	Front	3/4 × 4 × 35 1/4	1
Drawers			
T	Side	1/2 × 4 × 6	6
U	Front	1/2 × 4 × 10 1/4	3
V	Back	1/2 × 3 1/2 × 10 1/4	3
W	Face	1/2 × 4 1/2 × 11 1/4	3
X	Bottom	1/4 × 5 3/4 × 10 1/4	3
Hardware			
Y	Ceramic Knob	1 in. dia.	4
Z	Hinge	1 1/2 × 3	2
*Length includes tenons.			

member to keep the same face against the fence for all cuts, and to make a trial tenon and try it in the actual mortises.

Next, lay out the locations for the 3/8 in. pegs in the door joints. Make the holes in stiles (I) 1/32 in. farther from the shoulder than the holes in the tenon of the rails (J). That will help snug up the joint when you put it together.

Then cut the door panel (K) using an extension on your fence, as shown in Fig. 3. Set the table saw blade to 1 1/2 in. high and tilt it 16 degrees. The fence should be 3/16 in. from the blade and the blade must angle away from the fence. Make some test cuts in scrap to get the panel just right. Because of variations in wood thickness, as well as table saw accuracy, the degree of tilt should only be used as a guide. It's best to lay the panel dimensions out on a piece of scrap and set the saw to match.

When cutting the workpiece, clamp a feather board to the table so it hits the panel above the cut, as shown in Fig. 3. Cut across the grain first and then with the grain. That should minimize chip-out.

Cut the raised panel in the drawer faces (W) with the blade 1 in. high and inclined 19 degrees. Set the fence 1/4 in. from the cut. Again, make a test cut on scrap before risking your workpiece.

The raised panels stand 1/8 in. proud on both the door and drawers. The saw leaves an angled cut at the shoulder, so you'll have to square the cuts with a sanding block. Remember that the raised panel of the door floats in the frame so it is free to expand and contract with changes in humidity.

Next, dry-fit the door assembly, parts I, J and K. The panel shouldn't fit too snugly, and shouldn't extend to the bottom of the grooves because it needs room to expand. Add a dab of glue in the center at the top and bottom of the panel to equalize wood movement.

After gluing, let the pins protrude slightly and sand them flush later. Also, if you haven't done it before making the door, assemble the face parts of the base (C, D and E). Fair and sand the curved profile.

When gluing the parts, apply a little paste wax to the wood surface near the joints. Any glue squeeze-out won't penetrate the wood and you can clean off the wax with acetone or lacquer thinner.

The drawer faces are glued and screwed to simple drawer boxes as shown. Make the boxes to the actual openings. They are glued and nailed.

When assembling the dry sink, first glue up the base, then the upper section. Use screws to join all the pieces, even the drawer dividers. Countersink the screws and fill with 3/8 in. diameter birch dowels or pine plugs. Let them stand proud, and trim them flush after the glue dries. Glue and screw the facing stile (C) and rail (D) assembly after setting the back (G) into its rabbet. The back helps square the piece.

Note that you must insert the adjustable shelf (F) before closing in the front and back. The shelf won't fit through the finished opening. If, however, you want to retain the option of removing the adjustable shelf (F), then screw — but don't glue — the back (G) in place.

When gluing the upper section, only apply glue to the front 2 in. of the bottom (N). Join the base and upper section with the four cleats screwed and glued to the sides (A). Use slotted holes with ovalhead screws in the two rear cleats (M) so the bottom (N) can expand and contract. This assembly allows for wood movement, but directs any expansion toward the rear.

Sand and stain with Minwax Colonial Maple, and finish with two coats of penetrating oil. Finally, hang the door, mount the wooden door latch (L) with a 3/4 in. by no. 6 screw and add the ceramic knobs (Y) to the drawers and door.

OAK MAGAZINE RACK

This attractive magazine rack makes use of an especially strong joinery detail, the key spline. We decided to construct the project entirely of oak, although the key spline can also serve as a decorative detail if you make it with a contrasting wood such as walnut. The jig you'll need to cut the spline grooves is illustrated on page 64. We suggest that you use good material for the jig so it can be used in future projects utilizing the key spline.

Start by getting out sufficient stock for the sides (A), ends (B), stretcher (C), and handle (D). The construction of the rack is quite simple. The upper frame (consisting of two sides and two ends) and the lower frame (consisting of two sides, two ends, the bottoms (E), and the stretcher) are both assembled separately, machined for the key splines, and later joined with the 24 dowels (F).

Note that you'll need 2 in. wide stock for the handle, and that the ¼ in. radius roundover on the inside perimeter of the upper frame parts, and on the handle, must be established before the upper frame is assembled. Also note that you'll need to cut a ¼ in. by ¼ in. groove in the lower frame parts to accept the bottom panels. When cutting the miters on the sides and ends, it's best to use a stopblock for uniformity, and to start with these parts a little long so the miter cut establishes the length. Always test your miter gauge setting with some scrap stock before committing your project stock.

After gluing and assembling the upper and lower frames (don't forget to add the bottom panels, and to secure the stretcher with ⅛ in. by 1½ in. long dowel pins), you'll be ready to machine the key spline grooves. Refer to the illustration on page 64 for the construction of the jig that will hold the frames at the proper angle. As you'll see in the detail, only one setting of the fence is required to establish the ½ in. distance from the edge of the key spline groove to the edge of the stock. Rotate each frame to all four corners, then reverse the stock so the opposite face is against the jig to make the cuts for the second key spline. The handle is cut to length, but it is not shaped or added to the upper frame until after the key spline cuts have been made and the dowel holes have been drilled. The curved portion of the handle would interfere with using the jig for the second series of key spline cuts and with drilling the holes.

Now glue and insert the key splines. Note the grain direction of the splines is perpendicular to the miter for maximum strength. We recommend that you cut the splines slightly oversize, and then trim and sand them flush.

Once the key splines have been flushed with the frames, you can apply the ¼ in. radius roundover to the top perimeter of the upper frame, and the bottom perimeter of the lower frame. You can also lay out and drill the dowel holes in the lower frame, the upper frame, and the handle (not yet

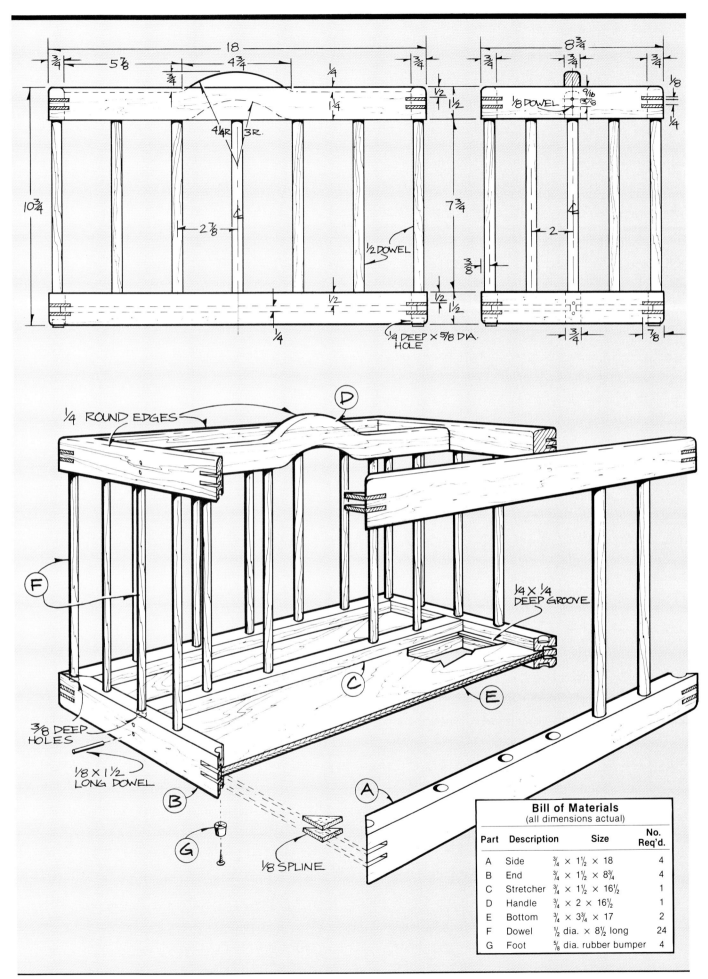

shaped or assembled to the upper frame). Having drilled these holes, now lay out and cut the curve of the handle, apply the radius, and assemble the handle to the upper frame using the 1/8 in. diameter dowel pins as shown. Also at this time, drill the 5/8 in. diameter holes to accommodate the rubber bumpers (G). We found the bumpers in a local hardware store, but you might want to check the size of the bumpers available locally before drilling the holes to accept them.

The oak dowel stock for the dowel parts can be ordered from The Woodworkers' Store, 21801 Industrial Blvd., Rogers, MN 55374. Since the dowel stock is sold in three-foot lengths and you'll obtain four dowel parts from each length, you'll need six lengths. The Woodworkers' Store also carries the dowel stock in cherry and walnut.

After final sanding all parts, we joined the upper and lower frames. Start by gluing and inserting all the dowels in the lower frame, then add the upper frame. Two coats of an aerosol spray satin lacquer, and the addition of four rubber bumper feet, completes the project.

KEY SPLINE JIG

As a functional joint and a decorative detail, the key spline is one of the most versatile, and easy-to-make joints in woodworking. We've used this particular joint in many projects over the years.

The jig used to cut the grooves to accept the key splines is just two pieces of wood that form a 90-degree angle. They are screwed to a back that will bear against an auxiliary fence on the table saw. Although the jig could be tailored to fit the project you are working on, we've found that the jig shown will serve adequately for most applications. Make sure that the 90-degree angle formed by the two pieces is accurate, and that this 90-degree angle is positioned 45 degrees from the surface of the table saw. Note that for the magazine rack, you'll need to set the saw blade for a 5/8 in. height, which will result in a 5/8 in. depth of cut. We used a blade that cut a 1/8 in. kerf so that each groove could be established with one pass. Since both splines are located 1/2 in. from an edge, there is no need to readjust the fence for the second spline cut; simply reverse the workpiece so the opposite side is against the jig. Although you could clamp the workpiece to the jig for each cut, we find that holding the workpiece firmly in place with one hand, while the other hand advances the jig, works perfectly well.

By altering the angle of the pieces holding the work, the key spline grooves can also be cut in various other multi-sided frames.

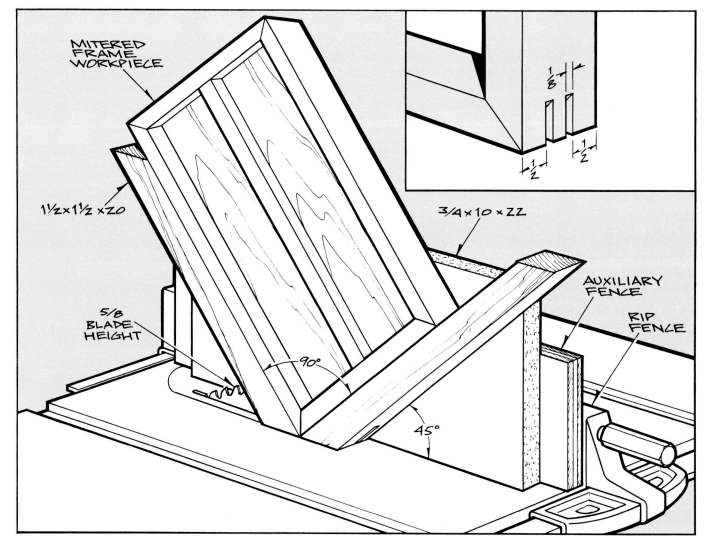

64 The Woodworker's Project Book

Shaker End Table

We discovered this lovely table in Hancock, Massachusetts at the Hancock Shaker Village *Shaker Workmanship '87* exhibit, where it was a juried selection. Although not a copy of any Shaker original, the table is certainly influenced by Shaker design.

Crafted in bird's-eye maple and cherry, this piece offers woodworkers an opportunity to practice a variety of woodworking techniques in a single project. The legs require turning, the aprons and stretchers involve mortise and tenon work, and the drawer features dovetailed joinery.

A good place to start is with the top (A). You'll need to edge-join several boards to get the required width. While you're waiting for the top to dry, go to work on the legs (B). We recommend that you start with turning blanks several inches longer than the finish length of the legs, and then trim the legs to final length after the various mortises have been cut, and the turning has been completed. This is important. If the legs were cut to final length before turning, mounting them in the lathe could easily cause the short section of end grain at the top of the mortises to break out. After cutting the 1 3/8 in. square blanks to rough length, lay out the various mortise locations. Note that the two back legs have identical apron mortises on adjacent sides, while the two front legs have the apron mortise and the stretcher mortises on adjacent sides.

There are a variety of ways to cut mortises. You could use the drill press to rough out the mortises, and then clean the sides with a chisel. Where multiple identical mortises are involved, though, as with this table, you

(continued on next page)

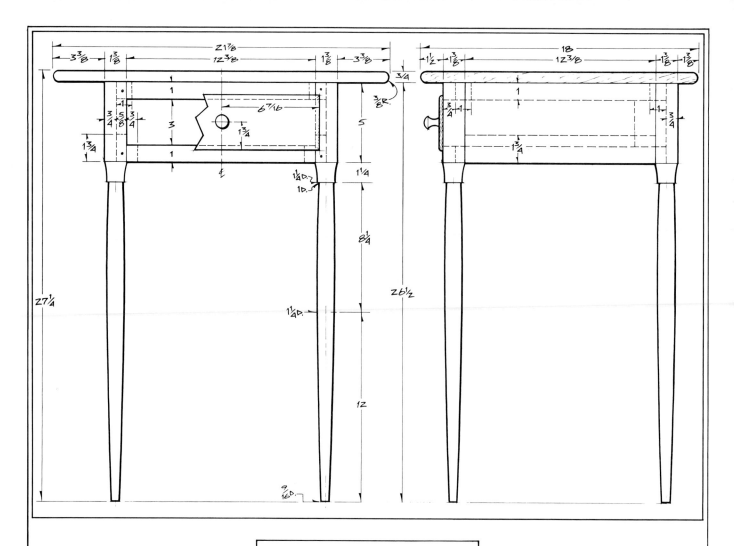

could also make templates and use the router to clean out the mortises. This latter method produces very clean sides and bottom, and leaves only a small amount of handwork with the chisel to square the mortise corners. You might choose to leave the corners of the mortises rounded, and then shape the corners of the tenons to match that radius. If you do use the router method, be sure to remove no more than about 1/8 in. of stock with each depth setting. With our 5/8 in. mortise depth, that would require five depth adjustments.

After the mortises have been established, mount the legs in the lathe with the top end in the live spur center, and the bottom end in the dead center. The legs are a basic spindle turning, accomplished with gouges and a skew. Sand the legs while they are still in the lathe, before removing them and crosscutting to final length.

Next, go to work on the stretchers (C), aprons (D), cleats (E), filler blocks (F), and drawer runners (G). Establish the tenons on the ends of the stretchers and aprons, referring to the appropriate tenon details. After assembling the table frame, cut the cleats, filler blocks, and drawer runners to final length and fit in place. Take special note of the slotted holes in the cleats. These are important since they will allow you to mount the top without restricting its wood movement in relation to the frame. Also note the inclusion of 3/16 in. by 1/2 in. dowel pins, which lock the stretcher tenons and add a nice visual detail.

The drawer is made last. Although designer Brian Braskie made the drawer in the traditional manner with handcut dovetails to join the sides, front, and back, we decided to illustrate the drawer with an applied face. By eliminating the half-blind dovetails at the drawer front, the dovetail work is made a little easier. To simplify the construction of the drawer case, we've made the ends (H) and sides (I) all the same width (3 in.). A 1/4 in. by 1/4 in. groove in these parts is cut to house the plywood bottom (J). The drawer face (K) is made separately, then glued to the drawer case. Since you've already used the lathe to turn the legs, you could also use it to turn the drawer knob (L). The dimensions you'll need for the knob are shown in the drawer detail.

After final sanding, you're ready for the finish. We think a piece as fine as this looks best with a natural finish that imparts a subtle satin gloss to the wood. For these qualities, we recommend a tung oil. Use two coats and rub out carefully with steel wool after both coats to achieve the desired luster.

Bill of Materials
(all dimensions actual)

Part	Description	Size	No. Req'd.
A	Top	3/4 × 18 × 21 7/8	1
B	Leg	1 3/8 × 1 3/8 × 26 1/2	4
C	Stretcher	3/4 × 1 × 13 5/8 *	2
D	Apron	3/4 × 5 × 13 5/8 *	3
E	Cleat	1 × 1 × 12 3/8	4
F	Filler Block	5/8 × 1 3/4 × 12 3/8	2
G	Drawer Runner	3/4 × 1 × 13 5/8	2
H	Drawer End	1/2 × 3 × 12 3/8	2
I	Drawer Side	1/2 × 3 × 12 3/8	2
J	Drawer Bottom	1/4 × 11 3/4 × 11 3/4	1
K	Drawer Face	1/4 × 3 1/2 × 12 7/8	1
L	Knob	7/8 in. diameter	1

*Length includes tenons.

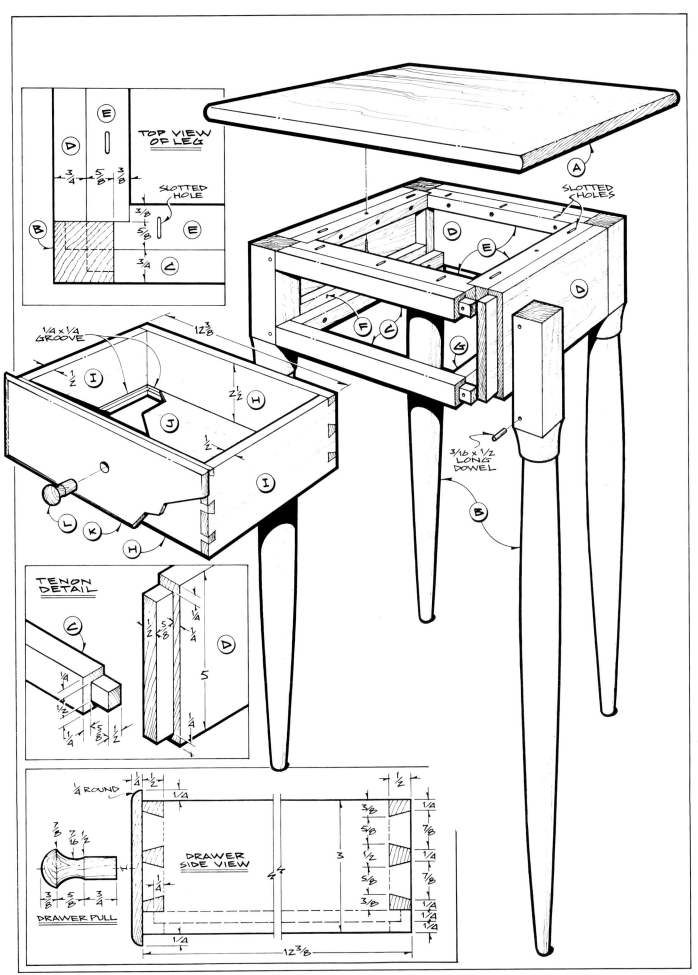

Heart Stool

This sturdy stool makes an ideal perch for your toddler, or a pair of tired feet.

It's made from 5/4 pine, and has angled legs for extra stability.

To start, cut the top (A), legs (B) and stretcher (C) to size. Add an extra 1 in. to the length of the legs to allow for the angled cuts you'll make next. Set the saw blade to 15 degrees, and make the cuts on the top and bottom of the legs. Use a good crosscut blade for a clean cut.

For the next operation — the slot at the top of the legs — we used a simple jig for accuracy. It's just a beveled block screwed to the miter gauge fence as shown in Figs. 1A and 1B. Note that the 2 1/4 in. slot is too deep for a standard dado cutter, so you'll have to use a saw blade for the cuts. A rip blade works best here.

The procedure is: set a stopblock on the miter gauge fence, and cut one side of the notch on both legs. Then reset the stopblock, and cut the other side of the notch on both legs. Keep the same edge of the workpiece against the stopblock for all the cuts. If you try to set up stopblocks on both sides of the workpiece, slight variations in the width of the legs can alter the width of the notches. After establishing the sides of the notches, remove the rest of the waste with repeated passes of the saw blade.

Next, cut the taper on the side of the legs with a band saw or jigsaw. Clean up that cut with a hand plane. Then transfer the heart pattern to the legs, and cut out the profile with a band saw or jigsaw. Also cut the 4 3/4 in. radius on the bottom edge of the stretcher. Clean up the heart and radius with files or a drum sander.

Next, lay out the matching 1/8 in. deep dadoes in the stretcher. Because the angles here can be confusing, it's a good idea to locate the sides of the cut using a registration line on the top surface of the stretcher (Fig. 2, Line A). Measure in 1 3/8 in. from each end of the stretcher and draw Line A across the top. Then place scrap at the 75-degree angle and mark Line B, as well as the cutting lines along the side of the scrap block. Also note that the scrap block must be the same 1 1/8 in. thickness as the legs. Repeat the marking for all four dado cuts.

When cutting the dadoes, be sure to set the miter gauge accurately. The setting is critical here because you have to reverse the miter gauge for two of the dado cuts. Since the back-to-back dadoes are cut at opposite miter gauge settings, any variation can prevent the legs from sliding into the stretcher.

Now, establish the 1/2 in. radii on the underside ends of the stretcher, and form the 3/4 in. radii on the corners of the top. Soften the edges of the top and stretcher, and counterbore 3/8 in. diameter screw holes in the stretcher.

Sand all parts thoroughly, and dry assemble the stool to make sure everything fits. Then glue the legs to the stretcher. The screws hold the top in place.

We painted the legs and stretcher with Soldier Blue Milk Paint made by The Old Fashioned Milk Paint Co., P.O. Box 222, Groton, MA 01450. For the top, we used four coats of tung oil.

Finally, countersink for the 3/4 in. diameter rubber bumpers, letting them stand 1/8 in. proud. The bumpers are available at most hardware stores.

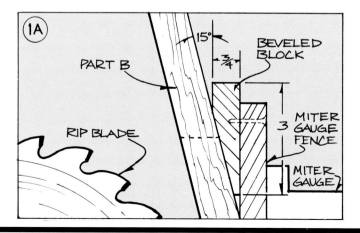

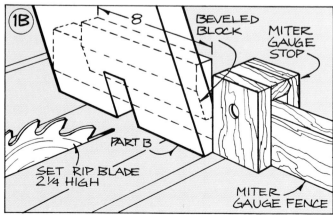

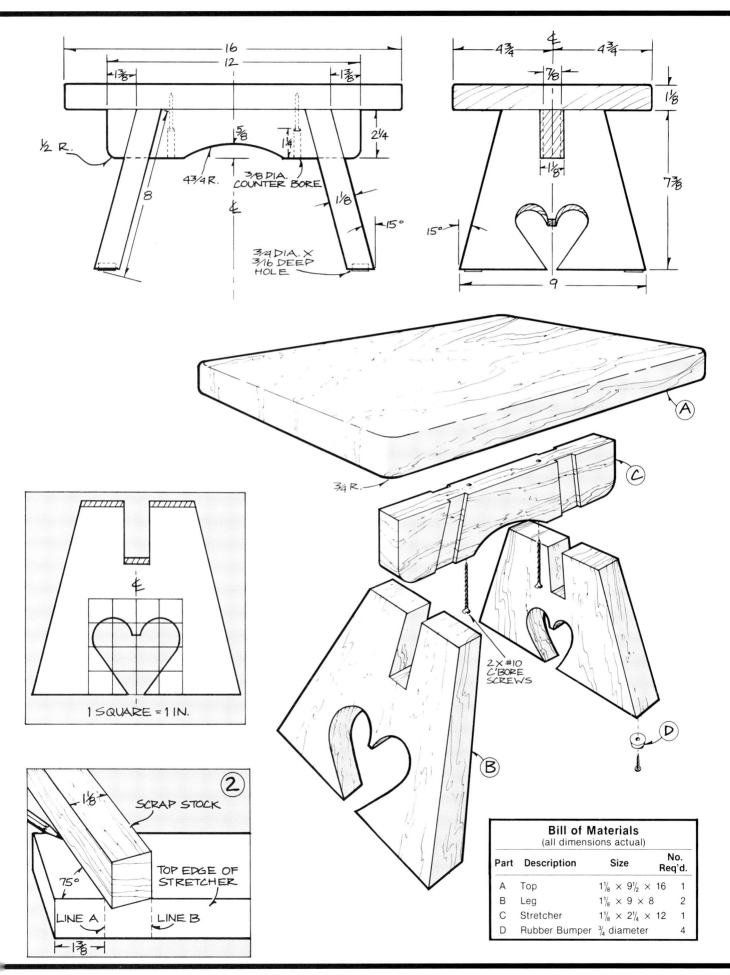

Decorative Cutting Boards

With cutting boards this attractive, you may be loathe to actually put a knife to them. In which case, you can always call them serving trays.

We show you how to make two different geometric patterns — one we call pattern A, the other pattern B. In the photo, Pattern A is being used to serve cheese and crackers, while Pattern B is standing on edge. Both patterns are easy to create with a table saw and miter gauge. You essentially just laminate a board containing contrasting woods, cut off pieces at a 60-degree angle, and glue them back together so the angled sections meet in various arrangements.

For our cutting boards we used walnut, satinwood and bloodwood, arranged as shown in Step 1. The stock, a full 1 in. thick, is available from Berea Hardwoods Co., 125 Jacqueline Drive, Berea, OH 44017.

Start by ripping the pieces to the sizes shown in Step 1 and joint the edges smooth. Then, depending on whether you want Pattern A or Pattern B, arrange the strips in one of the sequences shown in Step 1. Stagger them at a 60-degree angle and scribe an index line (Step 2). You'll use the index to line the strips up when you glue and clamp them.

For the gluing, use two sturdy clamp blocks (Step 3) and four clamps. Also use waxed blocks on top, as shown, to stop the strips from buckling under clamping pressure. Use a water-resistant glue, such as a plastic resin glue.

After the glue dries, set the miter gauge for the 60-degree angle. For a clean cut you'll need a good crosscut blade. When cutting off the strips, use an angled stopblock clamped to the miter gauge fence (Step 4). Also use a push stick to hold the strips as you cut. A word of caution: when cutting, don't back the workpiece up; the blade can catch the workpiece and throw it in the air.

Next, arrange the pieces as you'd like them for the cutting board (Step 5). For Pattern A, flip the pieces side-to-side. For Pattern B flip the pieces end-over-end. Label the strips so you can orient them correctly. Now, cut the ¼ in. wide spline grooves (Step 6). You'll need to make three passes with the router to reach the ⅜ in. depth. Note that the two end strips only need splines on one side, because the other side shows in the finished piece.

Cut the splines with a table saw, keeping them a little shorter and narrower than the ¾ in. groove-to-groove dimension. If they're too long or wide, they'll bind when you clamp the strips.

Next, apply glue to the splines and grooves, and glue up the tray. Clamp lightly side-to-side along the protruding points (Step 7) to help line up the pieces, then apply firm clamp pressure end-to-end. Putting light pressure on the points, which you saw off later, should make the lines of contrasting wood meet evenly. If not, make sure the splines aren't binding.

After the glue dries, sand off the excess adhesive. Then use the table saw to rip off the jagged edge on the sides, making a rectangular board. You'll next glue an edging strip to the sides, so you'll want a straight, square and smooth edge. To get that true edge, you can use a very fine crosscut blade.

Next, glue on the edging side strips (Step 8), which are cut a little wider and longer than shown. Sand or plane them flush after the glue dries.

Finally, sand and finish with Behlen's Salad Bowl finish.

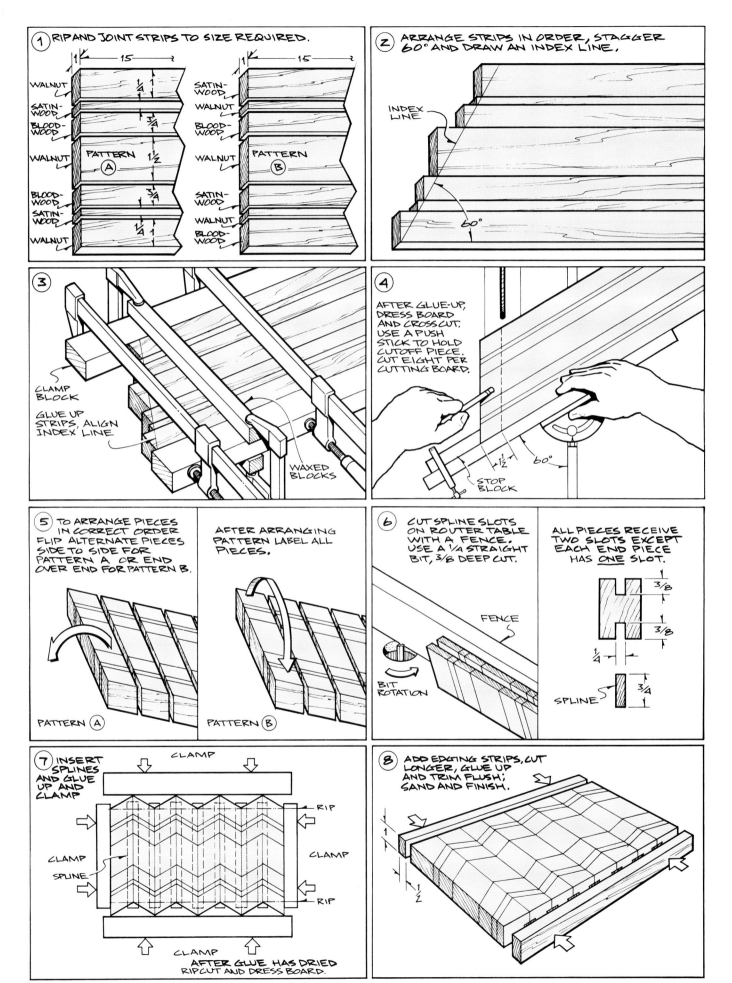

Kids' Piggy Bank

This little coin bank is easy to build, functional, and makes a great gift. It's also an ideal scrapwood project. We used two common hardwoods, maple for the center and walnut for the sides, but you can use just about any combination of woods. We do recommend that you select contrasting woods, and keep the dark wood on the outside and the lighter one on the inside, since the design seems to work best this way.

After sizing the stock (Step 1), lay out the coin cavity profile (see full-size pattern) on the center blank (Step 2). You won't need to trace the pig profile until later, after the sides have been glued to the center. Bore a starter hole and use the jigsaw or scroll saw to cut out the coin cavity (Step 3). Next, glue up the two sides around the center (Step 4). The extra length and width allowed in the side stock will be helpful here since there will probably be a little slippage as you apply clamp pressure. Now move to the router table, and use a laminate flush trimming bit to trim the sides flush with the center (Step 5). Switch to a $3/16$ in. straight cutter in the router table, and set up stopblocks as shown in Step 6 to rout the 1 in. long coin slot. You'll need at least three depth adjustments to cut the slot depth through to the coin cavity.

Use a $1\frac{1}{2}$ in. diameter Forstner bit in the drill press to cut the 1 in. deep recess hole that will accommodate the cover plate (Step 7). Switch to a 1 in. diameter Forstner bit and bore through to the coin cavity (Step 8).

After transferring the pig profile to the stock using our full-size pattern (Step 9), cut out the pig (Step 10). Smooth the profile with various fine mill files (Step 11), and then use a $3/16$ in. bearing-guided round-over bit to apply a radius to the edges (Step 12). A triangular mill file will final shape those areas where the round-over bit can't reach (Step 13). Next, establish the $3/16$ in. diameter eye and the $3/8$ in. diameter tail holes (Step 14).

The cover plate is fashioned as shown in Steps 15A and 15B, and fastened into place with two $1/2$ in. by no. 4 brass screws (Step 16).

After final sanding, we used an aerosol spray lacquer (Deft) to finish the piece.

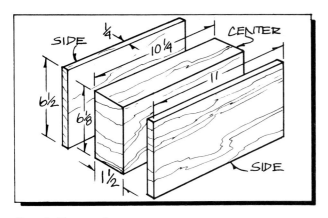

Step 1: Size stock

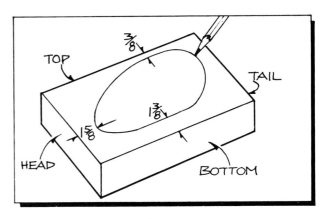

Step 2: Lay out coin cavity

Step 3: Cut out coin cavity

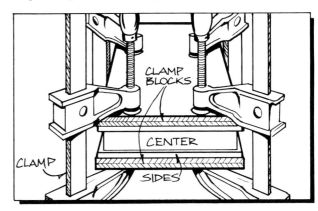

Step 4: Glue up sides and center

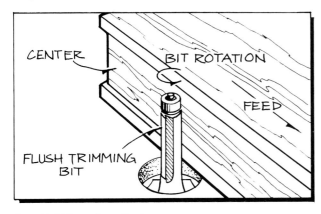

Step 5: Trim edges flush

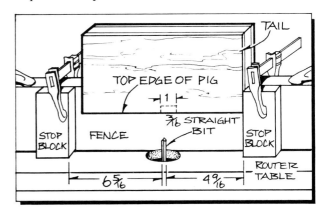

Step 6: Rout coin slot

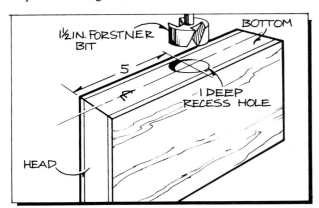

Step 7: Bore cover plate recess

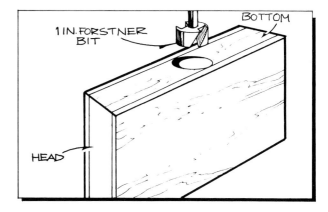

Step 8: Bore through to coin cavity

(continued on next page)

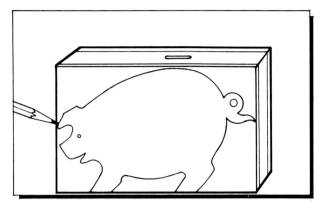

Step 9: Lay out pig profile

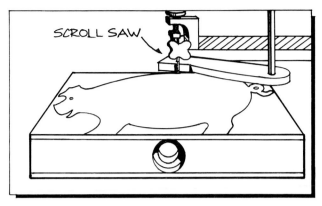

Step 10: Saw profile

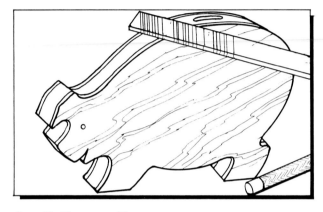

Step 11: Shape profile

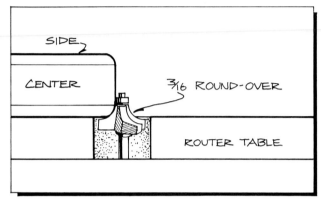

Step 12: Round edges

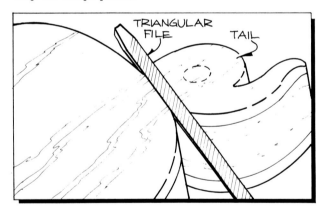

Step 13: Final shape

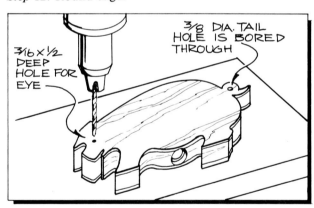

Step 14: Drill eye and tail holes

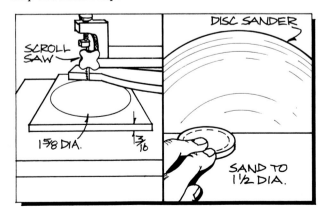

Step 15: Cut and sand cover plate

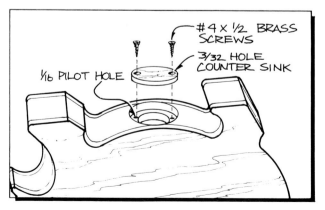

Step 16: Drill and mount cover plate

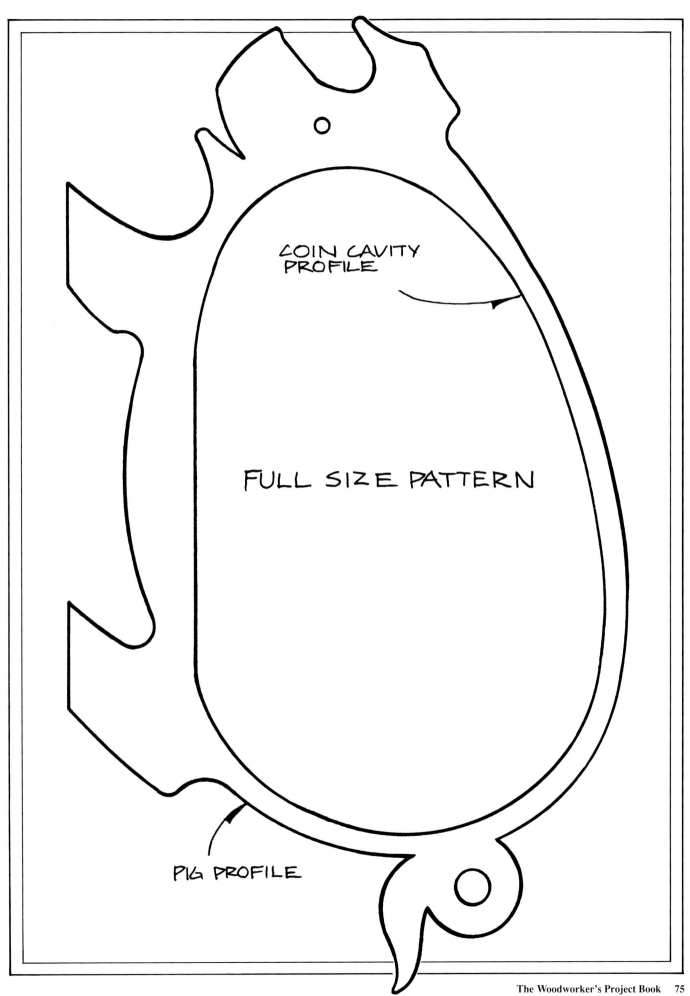

Turned Bowl

The classic turning details of our bowl give it a distinguished look, and provide plenty of practice for a novice woodturner. We chose walnut for its interesting grain, but many other hardwoods would also work.

We glued up three blocks of walnut to cut down on waste. You'll need a 1¾ in. thick block 4½ in. square, a 1½ in. thick block 6 in. square, and a 1½ in. thick block 9 in. square. Find the centers of the three blocks, then scribe a 4½ in. diameter circle on the 4½ in. square, a 6 in. diameter circle on the 6 in. square, and a 9 in. diameter circle on the 9 in. square. Cut out the circles with the band saw and glue them together. Keep the grain parallel. For centering, it helps to mark the diameter of the smaller discs on the faces of the larger disc. Note that for a strong glue bond, the faces must be flat.

To duplicate the bowl, transfer the full-size pattern to plywood or hardboard and cut it out to make two templates, one for the outside profile and one for the inside profile (Fig. 1). It helps to study the profiles for a few moments to get a visual image of what your hands will accomplish. You can the lathe for instant visual comparison.

There is waste stock on the turning block, so when marking the finished dimensions onto the workpiece, keep most of the waste on the headstock side. Use the outside template to mark the edges of the coves and shoulders.

Next, mount the workpiece on the lathe. You can screw the workpiece directly to the faceplate or glue it to a scrap block, which is then screwed to the faceplate. We screwed the workpiece directly to the faceplate using ½ in. by no. 12 screws. Don't use longer screws since they may interfere with the parting cut made later.

After mounting the workpiece, set the lathe for its slowest speed, and adjust the tool rest just below center and as close as possible to the rim of the disc. Rotate the workpiece by hand to insure it doesn't rub against the tool rest. When starting the lathe, always stand off to the side.
also tack the profiles to the wall behind

The outside profile is roughed out with the gouge, and then shaped with the gouge and roundnose (Fig. 2). You'll probably have to use the roundnose in the undercut section of the bowl exterior.

Because rough shaping will eliminate the first set of lines, you'll need to re-establish the profile for the finish shaping. When close to the final shape, hold the template to the workpiece for more accurate comparison. (Always stop the lathe when placing the template against the workpiece.)

To smooth the profile, use a shearing motion of the gouge and a light scraping action with the roundnose.

One difficult area is the shoulder separating the cove from the rounded bowl curve. To cut a smooth surface we used a parting tool in a slicing manner. After getting the approximate shape, slide the tool in from the cove side with the handle held low and the edge cutting at a very acute angle. The wood should peel cleanly away and leave a smooth shoulder. Check the outside profile against the template (Fig. 3). Also make a shallow cut with the parting tool to index the bowl base.

Next, use a 1 in. Forstner bit to bore out the center of the interior. Make the hole ⅛ in. short of the finished bowl depth, which is 2¼ in.

Now adjust the tool rest across the

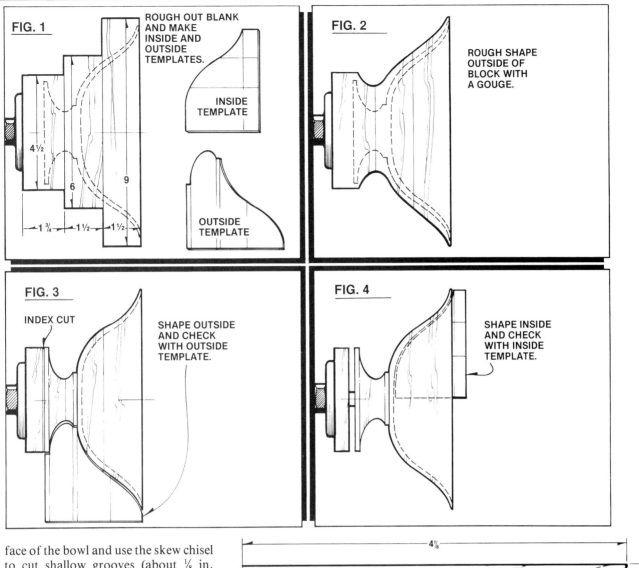

face of the bowl and use the skew chisel to cut shallow grooves (about 1/8 in. deep) in concentric circles about 1/2 in. apart.

When hollowing, start from the inside, closest to the 1 in. hole. You can hog out most of the material with the gouge, but you may need to use the roundnose for final shaping as well as for smoothing. As you get close to the final shape, use your fingers and the template to gauge the thickness of the bowl side (Fig. 4).

After shaping, remove the tool rest and sand with 150- and 220-grit paper. Then replace the tool rest and use the parting tool to establish the base (Fig. 4). Take care to stay well away from the screws used to mount the stock through the faceplate. Cut off the nub with the handsaw.

If you expect to put food in the bowl, use a non-toxic finish such as Behlen's Salad Bowl Finish. It's available from many of the woodworking mail-order suppliers, including Woodcraft Supply, P.O. Box 1686, Parkersburg, WV 26102.

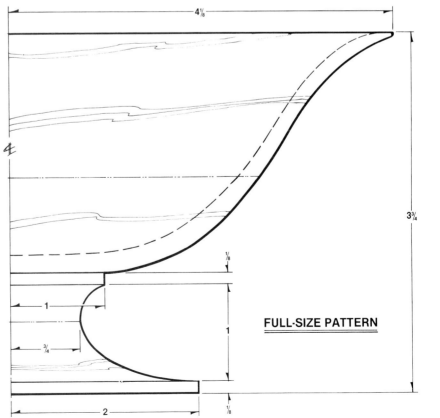

Country Cupboard

Now mostly replaced by kitchen cabinets, the standing cupboard was once found in nearly every home. And in some ways, they were more functional than modern kitchen cabinets. Frequently used tableware could be stored in a knife tray on the open top, with everyday dishes placed on the shelves. Moreover, the cupboard could be moved to any convenient spot, such as next to the kitchen table.

Rarely is a modern kitchen set up so that silverware and dishes are within a step or two of the kitchen table, where, after all, most of us usually take our meals.

This Country Cupboard can handsomely serve its original calling, or grace a special niche in the living room. Its open shelves offer an ideal spot for your collectibles, and its spacious lower cabinet provides plenty of storage.

Our cupboard is made with ¾ in. pine, available at any lumberyard. We used clear pine, but you may choose knotty pine for a more rustic look.

The top (A), sides (B), shelves (C) and door (D) are rather wide, so you'll need to edge-glue several boards. Make the panels slightly oversized, and use waxed blocks lightly clamped across the grain to keep the boards lined up when gluing.

While waiting for the glue to dry, you can cut the rest of the parts to size. You can also lay out the various radii on the top and bottom rails (F and G), and cut the profile with a jigsaw or band saw. Smooth the curves with a drum sander or wood files and sandpaper.

After the glue dries on the wider parts, flatten the surfaces with a thickness planer, a hand plane, or a belt sander. Then, cut the dadoes and rabbets in the top and sides. Also, lay out the radii and cut the profiles in the bottom end of the sides.

Next, sand all inside surfaces and assemble the sides to the shelves with screws and wood plugs. Place the top on the assembly and also attach that with screws and plugs. The back cleat (M) is glued and clamped to the underside of the top.

Next attach the bottom and top rails and the stiles (J). Use screws and plugs for the rails, but just glue for the stiles.

Attach the upper rail first, then the stiles, then the lower rail.

The tongue and groove back boards (H and I) are milled on the table saw and applied to the cabinet back with black drywall screws, as shown in the detail. Filler blocks (O) are glued in below the boards at either side.

To make the foot blocks (N), you'll need 1 in. thick stock. Drill holes for ⅞ in. diameter adjustable glides (R), which are available at most hardware stores. We often use a T-nut in place of the plastic insert, finding that the steel lasts longer. A dab of epoxy glue holds them in place. Use a V-block for clamping the blocks in place, as shown in the detail.

Next, lay out and rout the grooves in the stiles, as shown in the flute detail. Use the router with the edge-guide and a stopblock at the bottom and top of the cupboard. Use a ⅜ in. diameter core box bit and make several passes to reach the 3/16 in. depth. Work carefully. You only get one chance here.

For the doors, just cut the completed panels to the size of the actual opening and drill holes for the knobs. Use slotted screw holes in the door cleats (E) as shown. The slots allow the door panels to expand and contract with humidity changes. Now sand all remaining surfaces and stain the cupboard. We used Minwax Early American Wood Finish.

Take note that the molding (K and L), the door cleats and the back boards should not be assembled to the cabinet until after staining is complete. Leaving the pieces off until after finishing prevents wood movement from exposing bare wood.

The molding is shaped with a router and a classical bit. It's glued and tacked on with finish nails. On the sides, however, only use glue for about 3 in. toward the front. That will prevent shrinking or expanding wood from popping the molding loose. Although we shaped the molding ourselves (see detail), you could substitute a common bead or cove molding available at most lumberyards. The classical bit with a ¼ in. shank is available from Woodworker's Supply of New Mexico, 5604 Alameda N.E., Albuquerque, NM 87113. It's their part no. 800-903.

The hinges and knobs are available from Paxton Hardware Ltd., 7818 Bradshaw Rd., Upper Falls, MD 21156. Order part no. 5917 for the hinges, and part no. 5903 for the knobs.

Complete the project by attaching the hardware and coating the cupboard with several coats of a good quality penetrating oil.

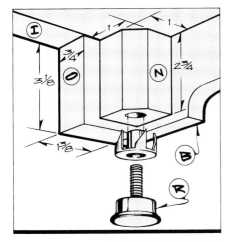

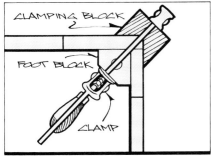

Bill of Materials
(all dimensions actual)

Part	Description	Size	No. Req'd.
A	Top	¾ × 15⅛ × 24½	1
B	Side	¾ × 13¼ × 59½	2
C	Shelf	¾ × 12½ × 21	4
D	Door	¾ × 8 × 21½	2
E	Door Cleat	¾ × 2½ × 6½	4
F	Bottom Rail	¾ × 3½ × 22	1
G	Top Rail	¾ × 5 × 22	1
H	Back (left)	¾ × 5⅝ × 56⅜	3
I	Back (right)	¾ × 5⅛ × 56⅜	1
J	Stile	¾ × 3 × 50¾	2
K	Front Molding	½ × ½ × 23	1
L	Side Molding	½ × ½ × 14½	2
M	Back Cleat	¾ × 1½ × 20½	1
N	Foot Block	1 × 1 × 2¾	4
O	Filler Block	¾ × 1⅜ × 3⅛	2
P	Knob	1 in. diameter	2
Q	Hinge	2 × 3	4
R	Glide	⅞ in. diameter	4

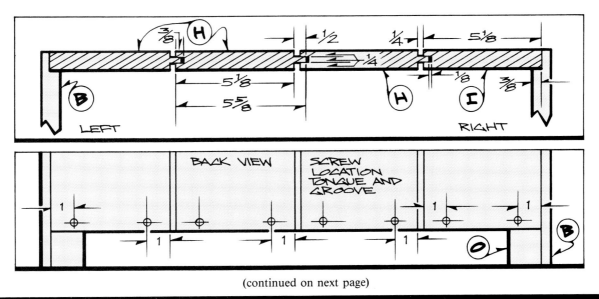

(continued on next page)

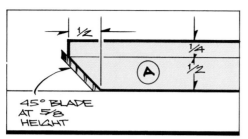
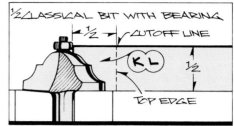
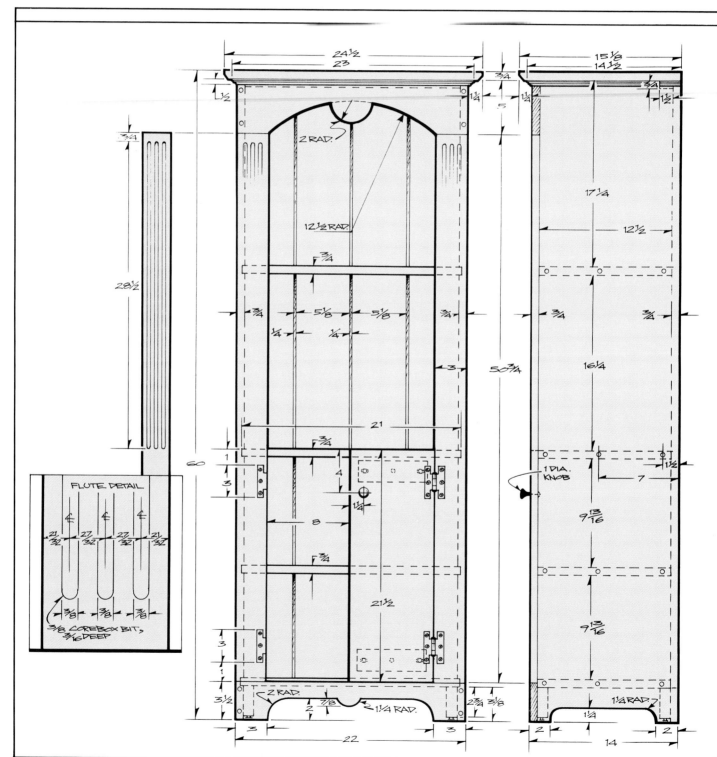

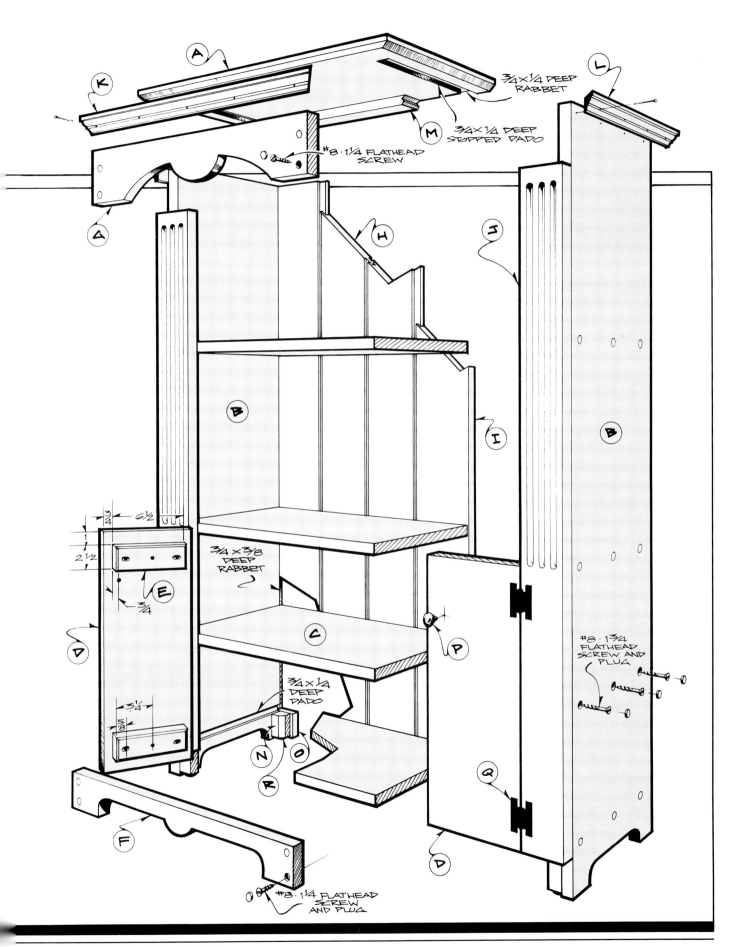

The Woodworker's Project Book

COUNTRY COFFEE TABLE

To the Shakers, function and form were as surely linked as the spoke and rim of a wheel. Every line of every board took its measure from its use.

Our Country Coffee Table would have been a puzzle to the Shakers. It's too short for a decent bench, too low to eat on, too big for a stool.

But the Shakers didn't sit on couches and they took no time for sipping coffee. If they had, they might have created this simple table, because its form also matches its function. The table top has ample room for coffee cups, nut bowls, and cheese platters. The large drawer easily holds the clutter swept off the top moments before the guests arrive.

Massachusetts woodworker Gene Cosloy designed and built the Country Coffee Table along traditional Shaker lines. He used cherry, but with walnut drawer pulls and pegs.

Start by joining several boards for the top (A). Leave them a little long and make the panel slightly wide. You'll trim it to exact size after the glue dries. Keep the boards aligned during glue-up by lightly clamping waxed blocks across the assembly. Another way to insure proper alignment is to use splines or dowels in the joints.

Next, cut stock to size for the front apron (B), the side aprons (C) and the rear apron (D), as well as the drawer guides (E) and drawer runners (F). Cut the 12 mounting blocks (G) from a ¾ in by 1⅝ in. strip with a rabbet notched into the side (See Detail). Make the strip about 14 in. long and crosscut it into the blocks. You can make the cutout for the drawers in the front apron with the hand-held jigsaw, or with hand saws.

Now, prepare the thick stock for the legs (H), truing and planing a plank before ripping it into four legs. Clean up the legs with a plane or a jointer.

Leave the lumber for the drawer parts off to the side. Drawers are always made to fit the actual opening.

Next, assemble all the parts the way you want them, paying close attention to the way the wood grains meet. When you have a pleasing arrangement, label the parts. Now, carefully lay out the location of the mortises and tenons, as well as the dadoes for the drawer guides, and the slot for the mounting blocks.

The mortises are open at the top, so you can cut them easily by hand. Use a dovetail saw to cut with the grain along each shoulder of the mortises. You'll only be able to cut a slanted kerf along the mortise shoulder (Fig. 1A). But the triangular cut serves as a guide for the chisel while removing the rest of the wood (Fig. 1B).

The tenons can also be cut by hand. First, use a handsaw to cut the $3/16$ in. deep shoulders on all the apron ends. Then cut into the end grain with a backsaw, carefully halving the line down to the shoulder (Fig. 2). If you're tentative about the cut, stay off the line and pare the tenon to size with a file. The tenon should slide easily into the mortise. If you have to force it in, it's too tight.

Next, drill the holes for the walnut pegs. Bore all the holes in the legs first, and then mark the matching holes in the tenons while dry-fitting the parts.

Note that the mortise in the legs only leaves a $3/16$ in. shoulder. The thin stock requires extra care when drilling the holes and clamping the legs to the aprons. You might want to drill for the pegs while dry-fitting the mortises. That way the tenon backs up the cut and prevents vibration while boring.

Cut the leg tapers next. Each leg has its two inside edges tapered from just below the apron. These can be cut on the table saw with a tapering jig (Fig. 3). But for the secondary taper toward the bottom, use a block plane. That taper, or chamfer, is cut along the inside corner. Lay out the taper, and cut down to the line with the plane (Fig. 4). Make sure you clamp the leg into the
(continued on next page)

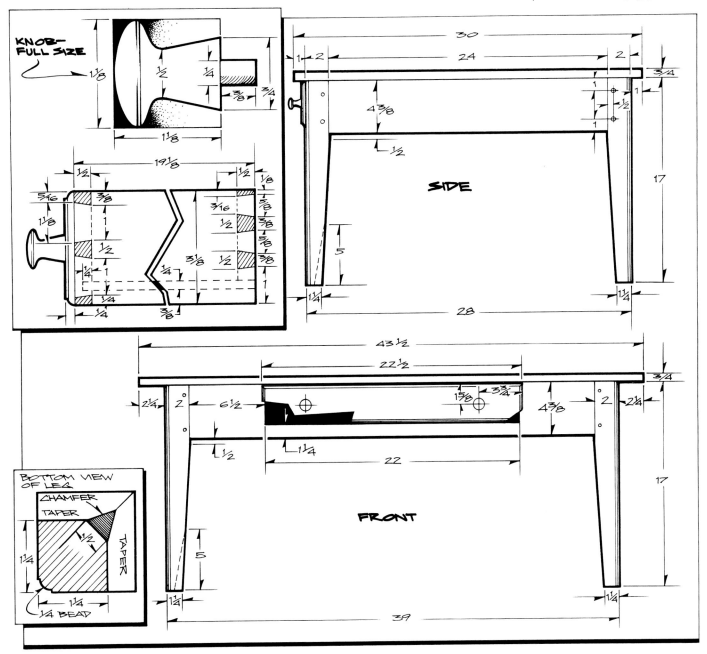

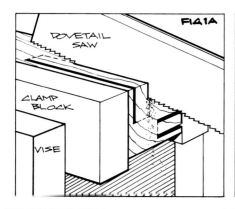
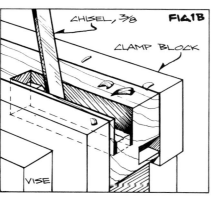
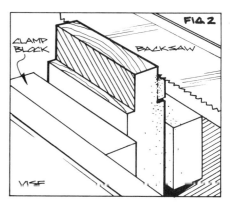

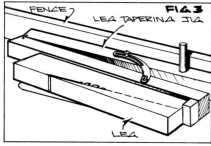
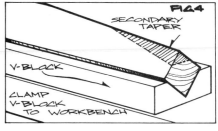

Bill of Materials
(all dimensions actual)

Part	Description	Size	No. Req'd.
A	Top	¾ × 30 × 43½	1
B	Front Apron	¾ × 4⅜ × 37	1
C	Side Apron	¾ × 4⅜ × 26	2
D	Rear Apron	¾ × 4⅜ × 37	1
E	Drawer Guide	¾ × 4⅜ × 27	2
F	Drawer Runner	¾ × ⅝ × 23½	2
G	Mounting Blocks	See Detail	12
H	Leg	2 × 2 × 17	4
I	Drawer Front	¾ × 3⅛ × 22½	1
J	Drawer Side	½ × 3⅛ × 19⅛	2
K	Drawer Back	½ × 2½ × 22	1
L	Drawer Bottom	¼ × 18⅞ × 21½	1
M	Drawer Pull	See Detail	2

V-block for stability.

Next, cut the beads into the outside corner of each leg. Use a ¼ in. beading bit in the router table.

Also using the router, cut the dadoes in the aprons for the drawer guides, and plough the grooves for the mounting blocks.

Now, assemble the table. When clamping the legs, be careful to apply pressure so each clamp pulls the tenon into the mortise. If the clamps squeeze the mortise, or apply cross pressure, they will distort the joint. It's best to work with two gluing sessions. First, glue the front and back aprons to the legs, and then glue the whole piece together. Use screws as shown to hold the drawer guides into the aprons. Glue the runners onto the guides.

The drawer is traditional dovetail construction. There are half-blind dovetails where the sides meet the front, and open dovetails where the sides meet the back. But you should note that the drawer front (I) only has a lip on the two sides. All four edges, however, have a ¼ in. bead detail.

To start, cut all the drawer parts except the bottom to size, then orient the parts and label them. Next, establish the ¼ in. deep lips on the sides of the drawer front.

Scribe the depth of cut for both tails and pins with a marking gauge. (The tails are the portion that resemble a dove's tail; the pins are the matching beveled shoulders.) Set the gauge a hair over ½ in., the thickness of the drawer sides (J). Scribe lines on the inside of the drawer front, on both ends of the drawer back (K), and on both ends of each side. When marking the sides and back, run the lines on all four surfaces. When marking the front, be sure to gauge off the inside of the lip, not the outside edge of the drawer.

After scribing the depths, mark the tails on the sides as shown in the dovetail layout. Extend those lines across the end grain. Use a sharp knife or similar implement for marking. Mark the waste sections with an "X."

Use a dovetail saw to cut the tails to depth. Stay on the waste side of the line. Use a chisel to cut along the depth line. But, for the first chisel cut, stay just off the line. The chisel creeps a bit toward the line when you first hit it with a mallet.

Next, trace the tail profiles onto the end grain of the drawer front and the back. Use a try square to run lines from the end grain to the depth marks.

Also use the dovetail saw and chisel to cut out the pins. For the half-blind dovetail in the front, you won't be able to cut all the way through with a saw. However, you can make a triangular cut from the inside. Hollow out the remainder of the cut with a chisel.

You may need some final chisel work to get the dovetails to fit correctly. They should go together snugly. If you have to bang on them, they'll split after gluing. The glue swells the wood.

After cutting the dovetails, make the ¼ in. deep by ¼ in. wide groove in the drawer sides and front to accept the bottom (L) and the ¼ in. bead on the drawer face. The drawer back fits on top of the bottom, which runs to the end of the side grooves. This arrangement allowed traditional drawer bottoms to expand toward the rear without distorting the drawer. You'll probably use plywood for the bottom, eliminating expansion problems, but it's still an elegant way to build a drawer. The bottom is cut after the drawer parts are dry-fit together to get the actual groove-to-groove dimensions.

Finish construction by gluing the drawer together and drilling holes for the pulls. You can either turn the pulls to the dimensions shown, or purchase similar ones. The pulls are available in maple from Woodcraft Supply, 210 Wood County Industrial Park, P.O. Box 1686, Parkersburg, WV 26102. Order part no. 50L41.

To finish the Country Coffee Table we recommend two coats of tung oil over a well-sanded surface. You should use at least 220-grit paper. The tung oil gives a nice close-to-the-wood finish, although it's not as durable as a varnish. Use a polyurethane varnish if you anticipate plenty of spills.

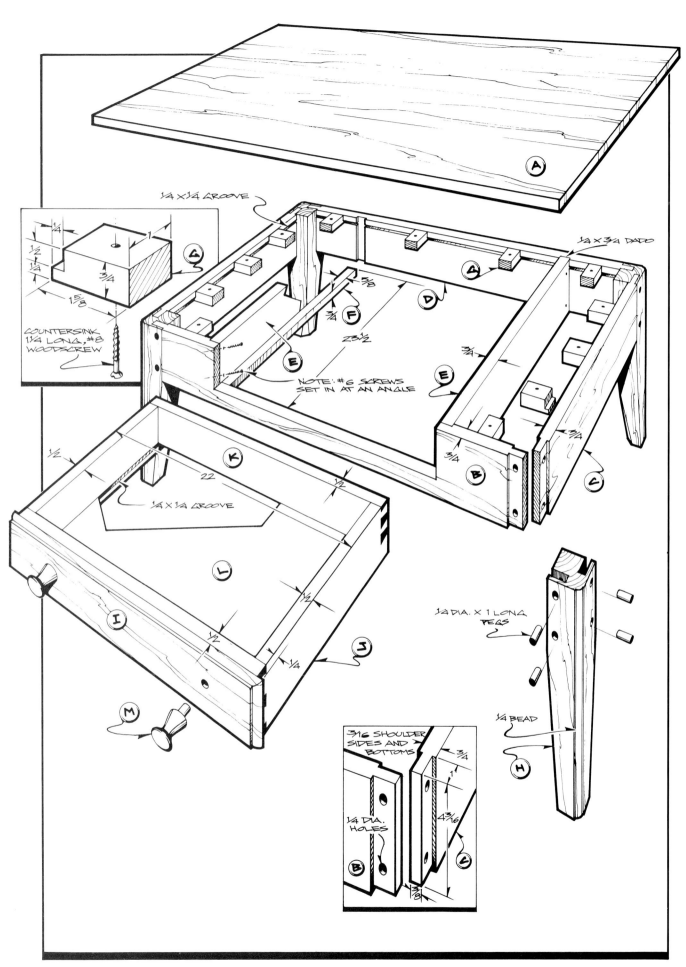

Our rooster adds a slice of barnyard to your windowsill, but comes without the predawn wake-up call. Don't be surprised, however, if the house cat resents the intruder on the perch.

We made the rooster from ¾ in. maple, but almost any wood will do. Softwoods work more easily, so you may choose to use pine, at least for the silhouette.

First transfer the full-size profile to the wood. Rough cut the rooster, preferably with a scroll saw. You can also use a band saw or jigsaw, although it will be harder to form all the facets. The inside details must be cut with a coping saw or scroll saw.

The most time-consuming part of the project is cleaning up the cuts. You'll need several shapes of small wood files to get into the corners. We used two sizes of half-round files, a round file, and a small triangular file. For sanding the edges, we used a dowel and some small slips of scrap wood wrapped with sandpaper to get into the facets.

Use a ½ in. half-round bit in the router table to cut the details on the base, which is built up with two ¾ in. boards.

We painted the silhouette with barn red milk paint. The paint, which covers in one coat, is available from The Old Fashioned Milk Paint Co., P.O. Box 222, Groton, MA 01450.

Rooster Folk-Art Silhouette

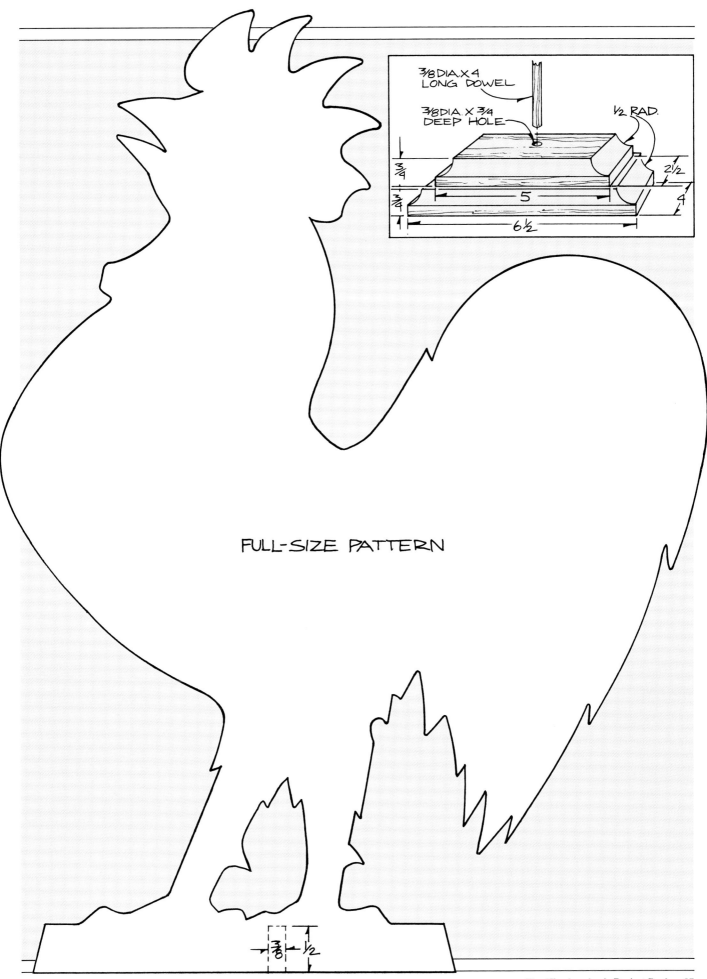

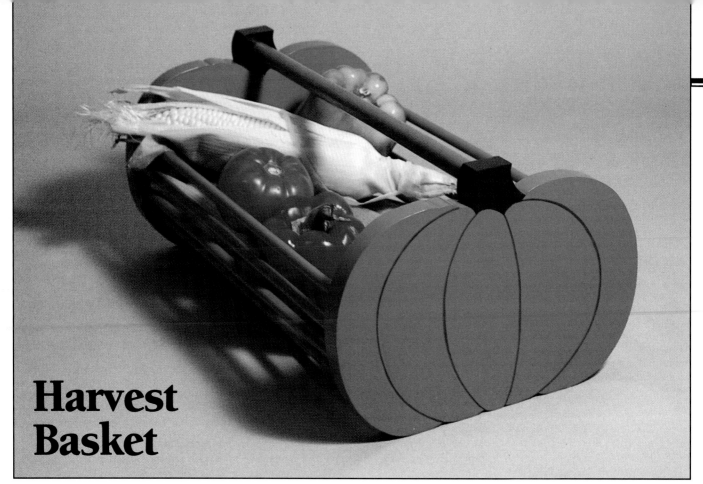

Harvest Basket

Those who have a backyard garden will find a basket like this handy for collecting the daily harvest. It's an easy weekend project that also doubles as a picnic basket.

We used ¾ in. pine for the pumpkin cutouts and birch dowels to connect them. The handle is ⅝ in. dowel, and the basket is ⅜ in. dowel.

The full-size pattern we provide makes cutting out the pumpkin shape easy. Just transfer the shape to tracing paper, mark it onto the wood and cut it out with a coping saw or band saw. Clean up the rough edges with wood files and sandpaper.

You'll need eight 36 in. lengths of ⅜ in. dowel and one length of ⅝ in. dowel. You can buy the dowels locally or order them from Craftsman Wood Service Co., 1735 W. Cortland Court, Addison, IL 60101.

We used oil paints on the pumpkin. First we applied a coat of flat paint to make a good base, then we used semi-gloss colors: orange for the pumpkin body, brown for the stem and green for the handle. Sand with 150-grit paper between coats of paint.

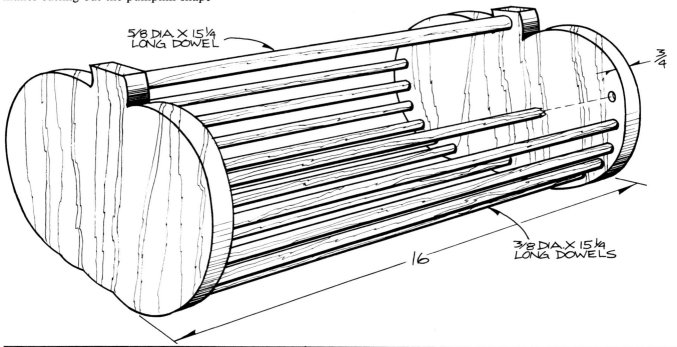

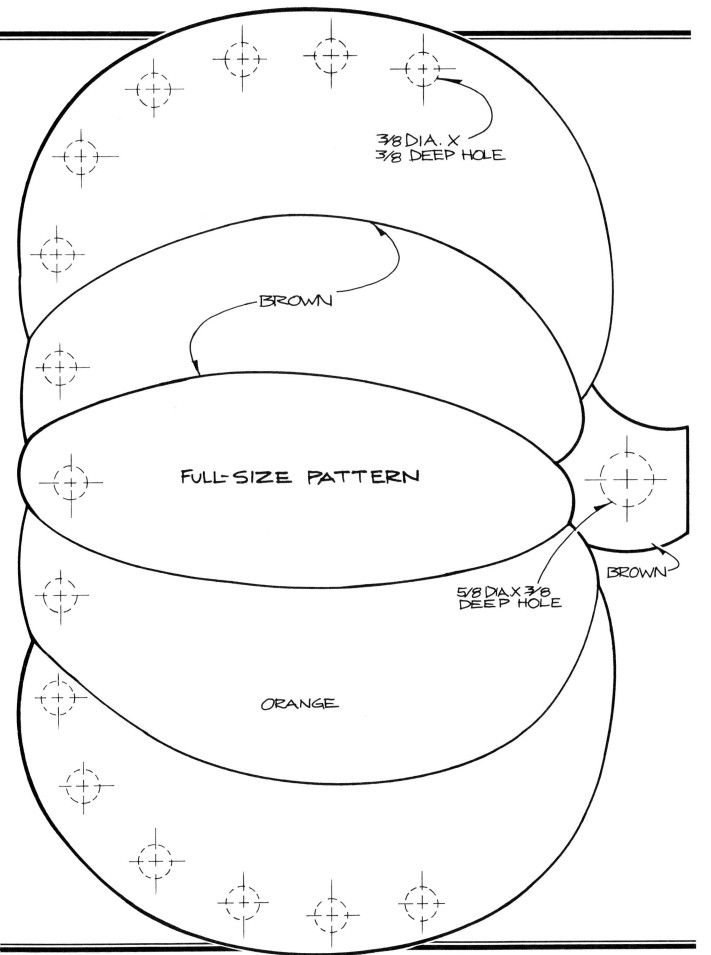

This bird is one of toymaker Skip Arthur's most popular designs, and is also one of the easiest to make. A cam on the wheels generates a flapping motion of the wings. Although he builds these birds of birch, with mahogany wings, you may want to try other woods.

The construction is very straightforward. After sizing the stock for the body (Step 1), you'll lay out the body profile, the eye, axle, wing screw holes, and the wing-pin slot (Step 2). Drill out the various holes and rout the wing-pin slot (Step 3) before you cut the body profile on the band saw (Step 4). The wing-pin slot is routed freehand and need not be precise, since it only provides a stop for the dowel pin mounted in the wing. Note that a scrap block protects the piece from being marred by the clamp. Next, round over the body profile using a 3/8 in. diameter round-over bit (Step 5). Note that the bit depth should be set slightly less than the full 3/8 in., so you'll have a small flat for the bearing to ride on.

Resaw stock for the wings using a pivot point guide on the band saw (Step 6). The pivot point enables you to accurately follow a line for resawing if your blade does not track absolutely parallel to the fence. After hand-planing the resawed stock for the wings, transfer the wing profile, the wing pin and the wing screw holes, and drill these holes out. Don't forget that you'll be working with mirror images on the wings, so get the wing-pin hole and the wing-screw countersink on the proper side of each wing relative to whether it is the left-side or right-side wing. Next, cut the wings on the band saw (Step 7).

You'll need 1¼ in. thick stock for the two wheels. Although you could lathe-turn the blanks for the wheels, it's just as easy to rough out the blanks with the band saw, and then sand them on the disk sander. Lay out the cam profile using the full-size pattern as a guide, and then use the dovetail saw and chisel to shape the cam (Step 8). Drill out the 5/16 in. diameter axle hole, and round the outside edge of the wheels (Step 9) before final assembling the bird (Step 10). Note that the dowel pin riding in the wing pin slot prevents the wings from pivoting too far. The wing movement depends on how you align the cams on the opposing wheels when the wheels are glued to the axle dowel. The cams aligned together will make the wings move in unison; 90-degree opposition of the cams will result in alternating wing movement. As with most wooden things for small children, no finish is always the best choice.

Bird Push Toy

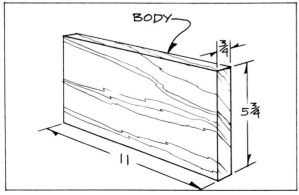

Step 1: Size stock for bird body.

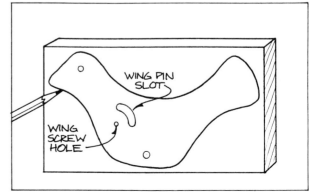

Step 2: Transfer bird profile, including eyehole, axle hole, wing screw hole, and wing-pin slot.

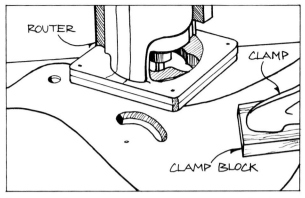

Step 3: Drill eye and axle holes; rout wing-pin slot.

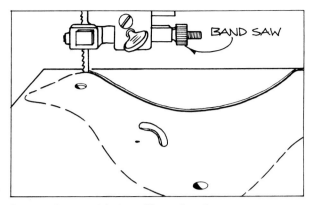

Step 4: Cut out bird profile on band saw.

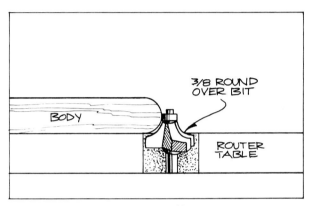

Step 5: Round over edges with ⅜ in. radius round-over bit in router table.

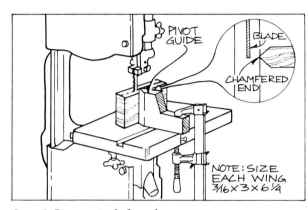

Step 6: Resaw stock for wings.

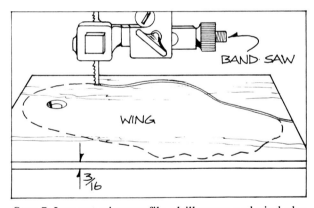

Step 7: Lay out wing profile, drill screw and pin holes in wing, then cut to shape on band saw.

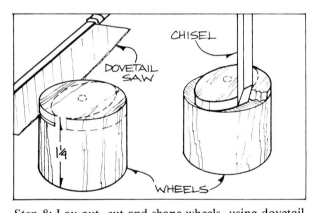

Step 8: Lay out, cut and shape wheels, using dovetail saw and chisel to shape cam.

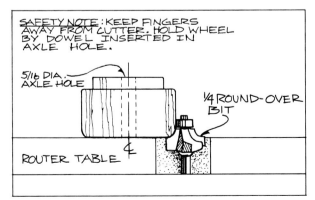

Step 9: Drill axle hole in wheels; round outside edge to ¼ in. radius.

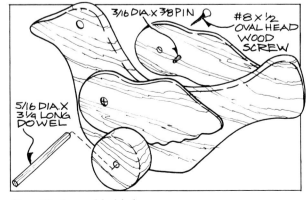

Step 10: Assemble bird.

(continued on next page)

The Woodworker's Project Book 91

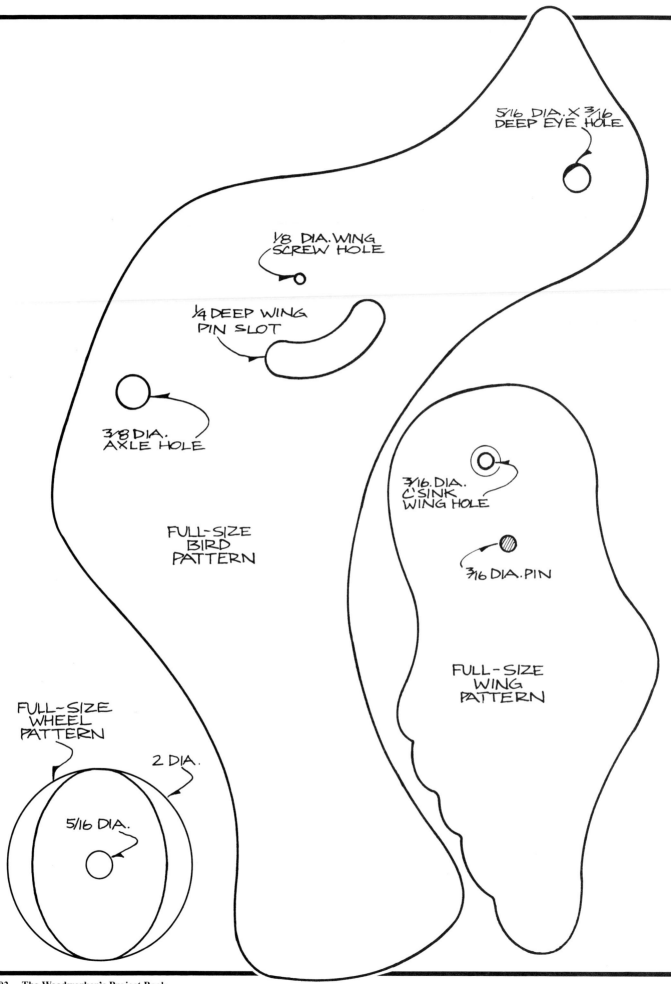

Pencil Post Nightstand

This handsome cherry nightstand makes use of classic pencil post legs, so you needn't have a lathe to make this project.

Begin by preparing and gluing up stock for the legs (A) and top (H). While you're waiting for these parts to dry, dress material for all the remaining parts, including the aprons (B, C, and D), cleats (E, I), drawer guides (F), backsplash (G), and drawer (J, K, L, and M). Note that the front apron part, long cleats, and drawer guides are oversize, since they are cut to fit later. Also, all the drawer parts should be oversize, since the drawer must be made to fit the actual opening after the stand has been constructed.

The front apron is cut from a single board. Rip a 1 in. wide strip, a 4 in. wide strip, and a 1 in. wide strip in that order, then crosscut the ends of the center section to create the drawer opening. If you are careful to match the grain when assembling the front apron, glue lines should be undetectable.

The mortises in the legs to accept the apron tenons are cut before the legs are tapered. Cut the matching tenons on the aprons, and test-assemble the legs and aprons. If everything fits, you can establish the leg tapers now.

To make the leg tapers, first lay out an octagon on the bottom end of the legs, and mark the point $22^{3}/_{4}$ in. from the bottom where the tapers start. Mark the

(continued on next page)

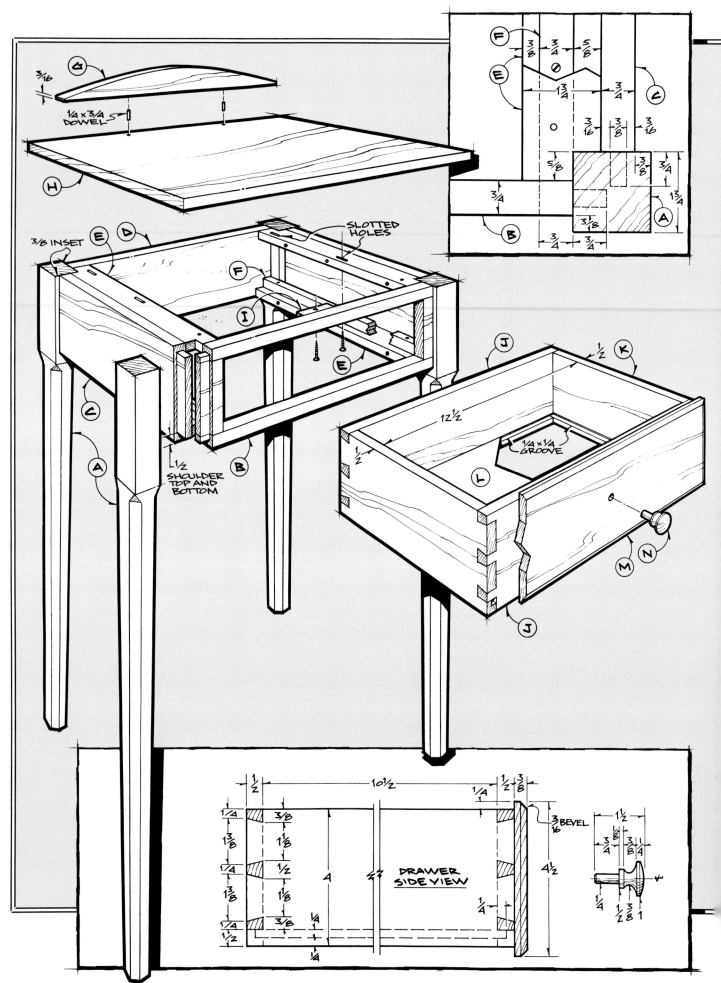

94 The Woodworker's Project Book

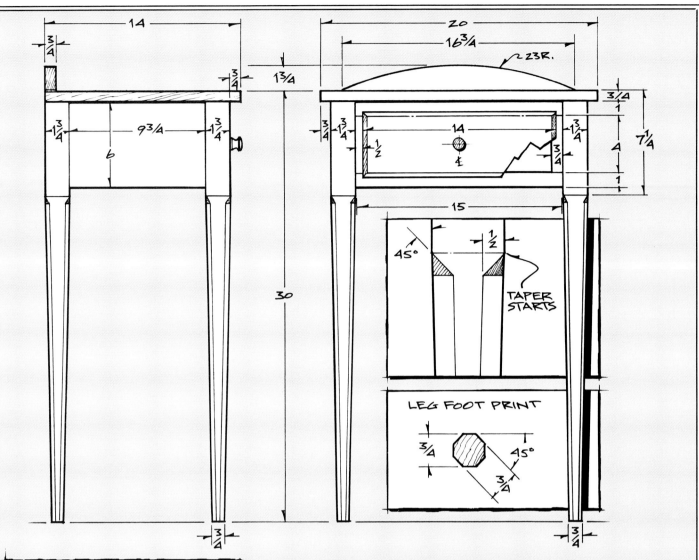

Bill of Materials
(all dimensions actual)

Part	Description	Size	No. Req'd.
A	Leg	1¾ × 1¾ × 29¼	4
B	Front Apron	¾ × 6* × 16½	1
C	Side Apron	¾ × 6 × 11¼	2
D	Back Apron	¾ × 6 × 16½	1
E	Long Cleat	1 × 1¾ × 11	4
F	Drawer Guide	¾ × ¾ × 11	2
G	Backsplash	¾ × 1¾ × 16¾	1
H	Top	¾ × 14 × 20	1
I	Short Cleat	¾ × ¾ × 3	1
J	Drawer End	½ × 4 × 13½	2
K	Drawer Side	½ × 4 × 11½	2
L	Drawer Bottom	¼ × 11 × 13	1
M	Drawer Face	⅜ × 4½ × 14	1
N	Knob	See Detail	1

*Width of stock for front apron must be wider before machining to allow for ripping cuts during apron assembly.

first four tapers, which are on the same plane as the four sides of each leg, and rough them in with the band saw. You'll need to shim under the end with one of the waste pieces to make the last band saw cut. Then use a sharp jack or jointing plane to smooth these four tapers to the line. The next four tapers are cut on the corners. Cut a pair of blocks with V-notches so you can clamp the legs on edge, then use a chisel to cut the bevel where the tapers start, and to establish the first several inches of each taper (rotate the leg for each taper). Then use the plane to establish the remaining length of the tapers.

Finish sand the legs and aprons before gluing and assembling them. This assembly should be done upside down and on a flat surface such as your saw table to keep the legs and aprons square.

Now cut and fit the long cleats and drawer guides. The slotted holes in these cleats will direct any movement in the top toward the back. Cut out the backsplash, and glue it to the top with several locating dowels, as shown. The dowels aren't needed for strength, but prevent the pieces from slipping as clamp pressure is applied. Add the short cleat along the front apron, and mount the top.

We chose to make a dovetailed drawer box with an applied face. The use of an applied face makes it easier to cut the dovetails. Note the chamfer around the face, softening the edge. You could lathe-turn the knob (N), or a similar knob, made of maple, can be ordered from Shaker Workshops, P.O. Box 1028, Concord, MA 01742. Specify their part no. 531.

Country
VEGETABLE BIN

The varnished top and punched tin panels add the right spice to this barn red vegetable bin. The potatoes, onions and summer squash won't taste any different, but the friendly cabinet may make you reach for the vegetables more often.

This small cabinet is also good practice in basic woodworking skills. The flush-mounted doors don't have lips to hide small size variations, so make sure to take extra care with the fit of the doors and cabinet. Also, the slip joints at the corners must be cut with precision for the doors to look right.

Begin by cutting the lumber to the sizes shown in the Bill of Materials. Cut all the parts except the top (A), back (E) and doors (G, H). You'll probably need to join boards for the top, and trim it to size afterward. The back is cut to fit after the cabinet is assembled. Doors are always made after the case is completed. With nominal 1 by 12 stock, you'll be able to get the 11 in. wide sides (B) and the 10¾ in. wide shelves (C) from single boards.

Next cut the radii on the sides and bottom rail (D), using a band saw or hand-held jigsaw. Then cut the grooves and dadoes in the sides, using a router or a dado blade. The dadoes across the grain are cut through for all three shelves. The rabbet for the bottom rail runs through to the shelf dado. And the rabbet for the back runs from top to bottom on both sides.

If you're using a router to cut the dadoes, clamp a straightedge across the sides to serve as a guide. It also helps to lay the two workpieces side-by-side and cut the dadoes in one router pass. Arrange the boards so that the two front edges butt together. That way, the router bit won't tear a chip from the finished edge.

After cutting the dadoes, mark the location of the screw holes as shown. Then drill and counterbore for the screws, which will be hidden by ⅜ in. diameter plugs. Before assembling the case, thoroughly sand all the parts you've made so far. It's easier to do the sanding now before the parts are assembled. Also dry-fit the parts to insure they fit, then measure for, and cut, the

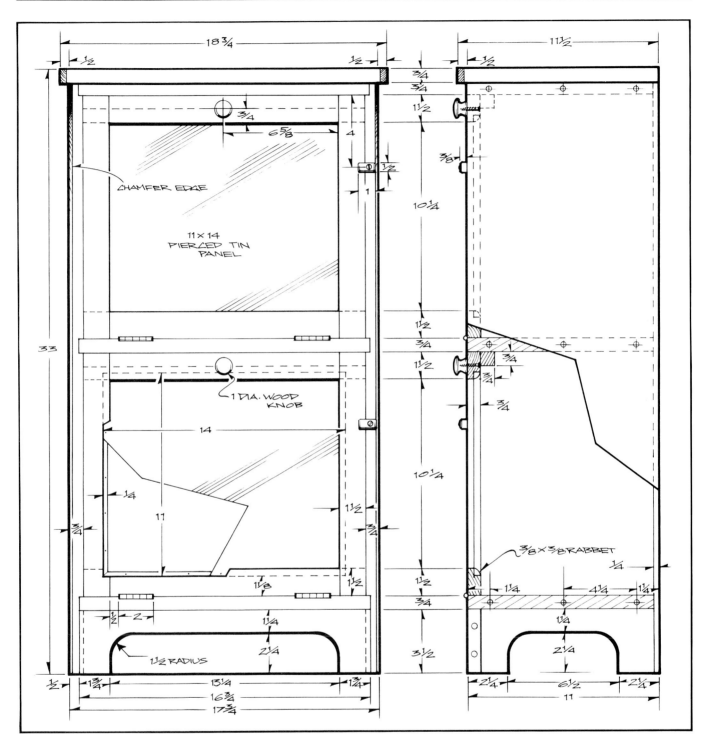

plywood back. When assembling, use glue sparingly to prevent it from squeezing out onto the wood surface. Pull the case together with screws and add the back, which should square up the piece. You don't need clamps as the screws will hold the shelves and bottom rail snugly in place while the glue dries. Before setting the case off to the side, carefully check it to make sure it's square. If the diagonals from corner to corner measure the same, then it's square.

Next, take the measurements for the doors, which should be about $1/16$ in. smaller than the opening. Carefully cut the stiles and rails to length trying for as clean a cut as possible. Pine chips easily, so sandwich the workpiece with scrap wood for each cut. A good crosscut blade is very helpful here.

For accurate slip joints it's best to use a table saw. Cut the mortised portion first and then size the tongue to fit. We used a tenon jig to cut both parts of the joint. The jig gives you added stability so you can safely cut the frame parts on end. Note that the initial shoulder cut for the tongue is cut with the miter gauge. Also note that you should always keep the same faces against the fence. Don't flip the pieces

(continued on next page)

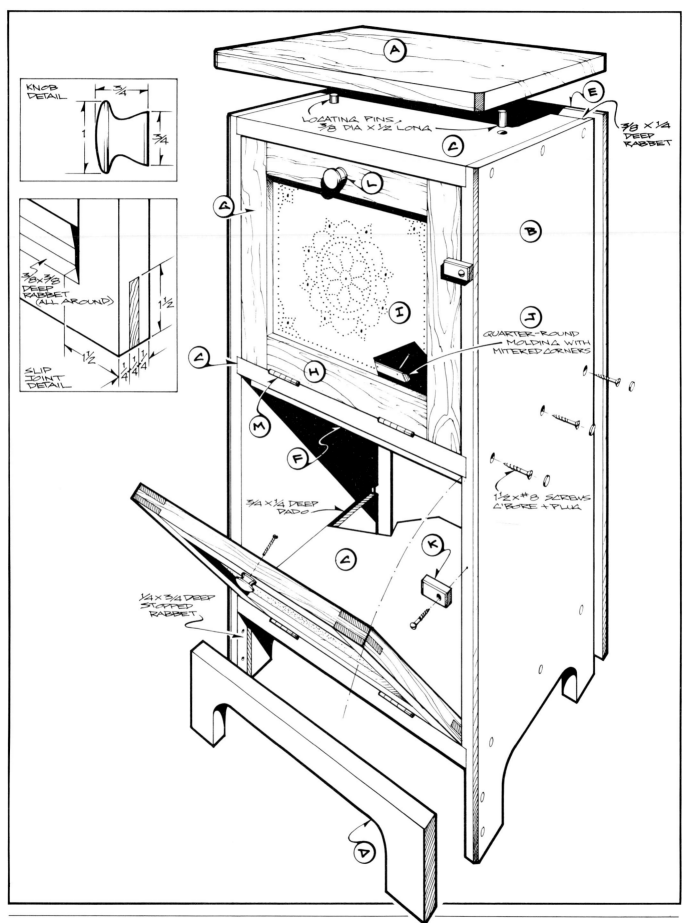

over for successive cuts; instead turn them end for end.

After gluing up the doors, cut the rabbet for the tin (I) with a router and a ⅜ in. rabbeting bit. Square off the rounded corners with a chisel. For the retainer strips (J), we used quarter-round molding.

Finally, chamfer the front edges of the cabinet with a hand plane, nip the corners of the top to match, and mortise for the hinges (M) in the doors and shelves. Note that for the doors to work smoothly, the hinges must be carefully mortised into the frames and the case. Now plug the screw holes with ⅜ in. diameter plugs. Cut the plugs off close to the cabinet after the glue dries. Thoroughly sand the cabinet and doors.

To punch the tin, make a template from our full-size pattern. The pattern is symmetrical so you can use the quarter-section we supply to make the whole template. Lay the paper template over the tin and punch through it to form the holes. Use a 1/32 in. nail set for the small holes, and a 3/32 in. nail set for the larger holes. Or, you can use regular tin punches if you have them. Be careful to keep all the holes uniform. We used Pie Safe Tin available from the mail-order company Country Accents, P.O. Box 437, Montoursville, PA 17754. Be careful handling the tin: fingerprints can etch it. For more on tin punching, refer to the Pierced Tin Step-by-Step Instructions on page 10 of this book.

Finish the top and doors separately from the cabinet so you don't get paint on the varnished surfaces. Screw or glue the top to the case after you've finished painting and varnishing. We used polyurethane for the top and doors and barn red milk paint for the case. The milk paint is available from The Old Fashioned Milk Paint Co., P.O. Box 222, Groton, MA 01450.

Finally, mount the doors and add the knobs (L) and turnbuttons (K). The turnbuttons are made from scrap, and screwed into the case with ½ in. by no. 4 screws. You can also add lid supports, available at most hardware stores, to keep the doors from resting on the knobs when open.

Bill of Materials
(all dimensions actual)

Part	Description	Size	No. Req'd.
A	Top	¾ × 11½ × 18¾	1
B	Side	¾ × 11 × 32¼	2
C	Shelf	¾ × 10¾ × 16¾	3
D	Bottom Rail	¾ × 3½ × 16¾	1
E	Back	¼ × 17 × 32¼	1
F	Door Stop	¾ × ¾ × 16¼	2
G	Door Stile	¾ × 1½ × 13¼	4
H	Door Rail	¾ × 1½ × 16¼	4
I	Door Panel	11 × 14 tin	2
J	Retainer	¼ in. quarter round	as req'd.
K	Turnbutton	⅜ × ½ × 1	2
L	Knob	see detail	2
M	Hinge	1½ × 2	4

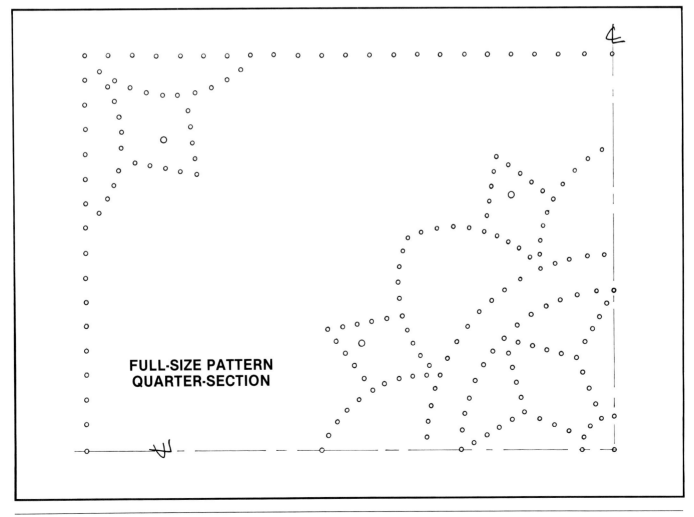

FULL-SIZE PATTERN QUARTER-SECTION

This delicate scroll saw project will add a nice touch to your collection of Christmas ornaments.

You can probably complete the project in an afternoon or two. And all you need for wood is a few scraps of pine and a short piece of leftover dowel stock.

Dress some pine to a 1/8 in. thickness. Use clear pine, as any knots will make the ornament too fragile. Then transfer the full-size pattern to paper. There are five parts. Make two copies of each of the two small half circles.

Attach the patterns to the pine with rubber cement (or spray adhesive) and carefully cut out the ornament parts with a scroll saw. Don't use a handheld jigsaw; the pieces are too delicate. Be as accurate as possible, especially on the tabs and cutouts where the sections join. After cutting them out, test fit the parts, and shape them as needed with files and sandpaper. Also sand off the paper and rubber cement from the workpieces.

When assembling, note that the circle halves without tabs meet the circle halves with tabs, and that the parts come together at the cutouts in the larger circle. Glue the ornament parts together using four thin rubber bands as clamps. We used rubber bands 2 1/2 in. long and ran them from point to point.

For the ornament holder use a 1/2 in. diameter by 2 in. long dowel. Use a V-block and hand screw to hold it steady on a drill press while boring the 5/16 in. hole in the center. The hole in the dowel is 1 1/2 in. deep. To round the end of the dowel, chuck it lightly in the drill press and use a sanding block against the turning workpiece.

To make the 3/8 in. deep kerf cut in the rounded portion, use the V-block again. Clamp the V-block and dowel to a miter gauge fence with a hand screw clamp, and cut the kerf with the table saw (See Detail). If your saw-blade kerf is less than 1/8 in., you may need two passes. Use a dab of glue to hold the ornament in the dowel holder.

You can either varnish the ornament or paint it. For the painted version, we used red enamel on the outside edges and white for a base coat. Note that you may need to trim the tree top spire to fit inside the holder.

Treetop Christmas Ornament

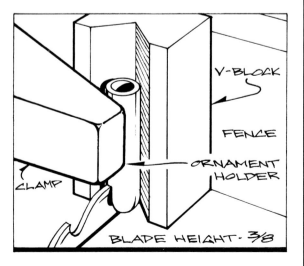

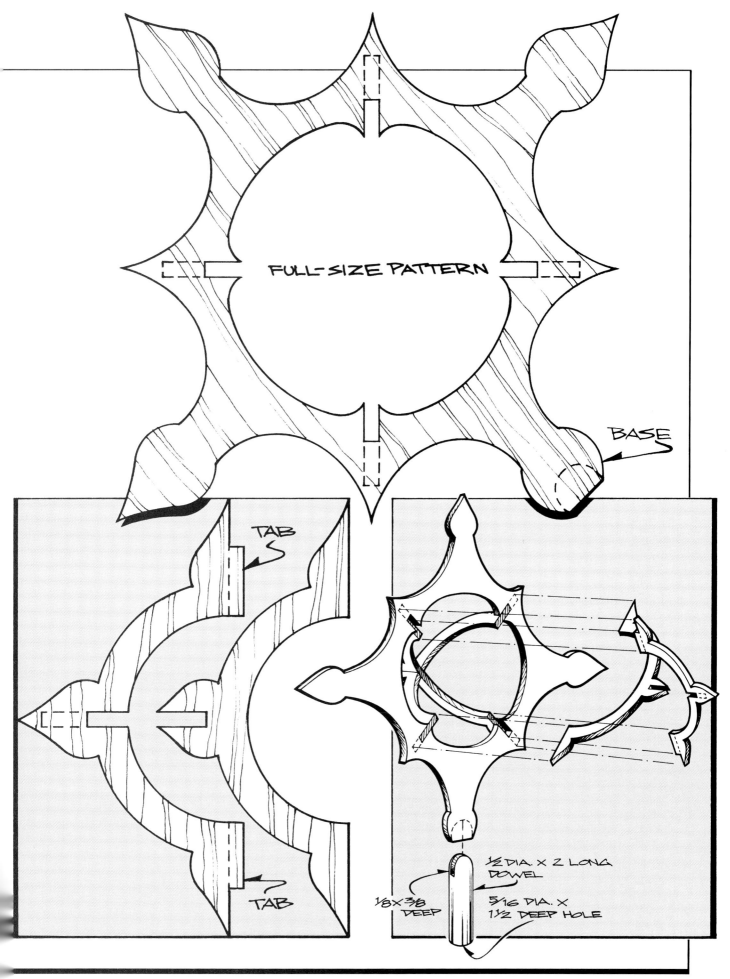

CLASSIC PICKUP TRUCK

We found this handsome pickup truck at the American Crafts Council (ACC) Fair in Springfield, Massachusetts. Designed and built by Kansas woodworkers Fred Cairns and Kathy Dawson, this truck is built to take a beating. Cairns said that to demonstrate their durability, he has actually stood on them.

The truck shown is constructed of oak and padauk. Although you could build it in pine, the truck will be more durable if you use hardwoods. The choice of the contrasting woods will make it a project you'll be proud to display on the desk or mantle.

Although not hard to build, details make the truck seem more complicated than it actually is. Start by getting out and cutting stock for parts A through L. Only the cab top (F), seat parts (I, J) and fenders (K) are padauk. The remaining parts are oak, except for the wheels, pegs, and dowels, which are maple or birch. We've arranged with a toy-parts supplier to provide a package kit that includes all the turned parts, except the gear shift, steering column,
and dowels. The kit does not include the wood you'll need to build the truck. Ordering information is listed in the Bill of Materials. Be sure to have the parts on hand before you start building.

After cutting the chassis (A), cab front (D), cab back (E), cab top (F), bed (G), bed side (H), and seat parts, apply a $\frac{3}{16}$ in. radius to the front end of the chassis. This will be a router table operation. Note that you'll also need to cut a bevel on the front end of the cab top. While you could establish this bevel with the table saw, it's easier to just take a small block plane and cut it by hand.

Next cut stock for the upper hood (B) and lower hood (C). After cutting these parts, use the table saw to establish the vertical louvers, as shown in the lower hood detail. First mark the location of the slots on either side of the lower hood. Then, with the table saw blade set for a $\frac{1}{8}$ in. depth of cut, remove the area between your marks. Now reset the blade for a $\frac{1}{4}$ in. deep cut, and establish the five slots. Just line the blade up with your marks for
each of the slots. You need not be too precise here. Use the miter gauge fence to guide the stock through the blade, and keep your hands up high, away from the blade. Assemble the upper hood to the lower hood, and round the upper hood using the band saw. After sanding, use the router table with a $\frac{3}{16}$ in. radius bearing-guided round-over bit to detail the front of the hood assembly.

Now join the chassis, hood assembly, cab front, cab back, cab top, bed and bed sides. The dowels that join the chassis, cab front, cab back, and cab top add some mechanical strength, and also serve to locate the parts during assembly. Note that with the hood, chassis, and cab front assembled — but before adding the cab back, cab top, and bed parts — you'll need to drill the holes for the foot pedals (P), steering column (T), and gear shift (U). It's important to drill these holes now, since there's no room for a drill inside the cab once the assembly is complete. Glue the steering wheel (O) and steering column, foot pedals, and gear shift in place. When dry, sand the truck chassis/hood/cab assembly flush on the sides. A disk sander, if you have one, comes in handy here.

Next, use the ³⁄₁₆ in. radius bearing-guided round-over bit, mounted in the router table as shown in Fig. 1 to establish the round-over detail. This detail runs around the cab and down the bed sides. Note that the bit height is set so that a small shoulder is established. The round-over detail stops short of the hood, and the chassis, and is flush with the bed. Use files and sandpaper to round the edges of the windshield and back window.

Now drill for the axle pegs (N), parking light pegs (Q), and radiator cap (S). Take extra care when drilling the holes for the axle pegs. Find the centerline of the chassis, mark their locations as shown on the side elevation, and use the drill press to get a good, straight hole. This way you'll be sure that all four wheels will be evenly positioned on the same plane. The bumper (L) should be in position when you drill for the front parking light pegs, which help secure it. Note the slight chamfer around the front edges of the bumper. Glue up the two seat parts, and set aside to dry.

To make the fenders, first lay out a ½ in. grid pattern on one of the two fender blanks. If you have access to a copy machine with the capacity to enlarge, this is a quick way to make a full-size pattern. Transfer the fender profile, then temporarily join the two ¾ in. thick fender blanks with hot glue, and cut out the fender profile with the band saw. Sand the fenders, then round the edges, either by hand or with a ⅛ in. radius round-over bit on the router table. Then separate the fenders and glue them in place on the truck.

Lastly, add the wheels, headlights, radiator cap, and seat. When inserting the axle pegs, use very little glue to avoid any glue squeezing out. It is difficult to remove excess glue from behind the wheels. The edges of the seat are rounded with the ⅛ in. round-over bit, or by hand, before it is mounted.

It's best to keep toys such as this away from children under 3 years old, since small parts that break off could present a choking hazard. We don't usually recommend a finish for toys, however a non-toxic salad bowl finish, such as Behlen's, will help to show off the character of the wood.

Bill of Materials
(all dimensions actual)

Part	Description	Size	No. Req'd.
A	Chassis	¾ × 3 × 11½	1
B	Upper Hood	¾ × 3 × 2¾	1
C	Lower Hood	1¼ × 3 × 2¾	1
D	Cab Front	¾ × 3 × 3½	1
E	Cab Back	¾ × 3 × 3½	1
F	Cab Top	½ × 3 × 4⅛	1
G	Bed	¾ × 3 × 5	1
H	Bed Side	½ × 1¼ × 5	2
I	Seat Bottom	¾ × 1 × 2½	1
J	Seat Back	½ × 2 × 2½	1
K	Fender	see pattern	2
L	Bumper	⅜ × ¾ × 4½	1
M	Wheel*	2 dia. × ¾ thick	5
N	Axle Peg*	see detail	4
O	Steering Wheel*	1 dia. × ⁵⁄₁₆ thick	1
P	Foot Pedal Peg*	see detail	2
Q	Parking Light Peg*	see detail	4
R	Headlight*	¾ dia. × ⁷⁄₃₂ long	2
S	Radiator Cap*	see detail	1
T	Steering Column	¼ dia. × 1¾ long	1
U	Gear Shift	³⁄₁₆ dia. × 1½ long	1

* These parts are included in kit. Order "Classic Pickup Truck" kit from Lynes Unlimited, Route 2, Greenleaf, KS 66943; tel. (913) 747-2612.

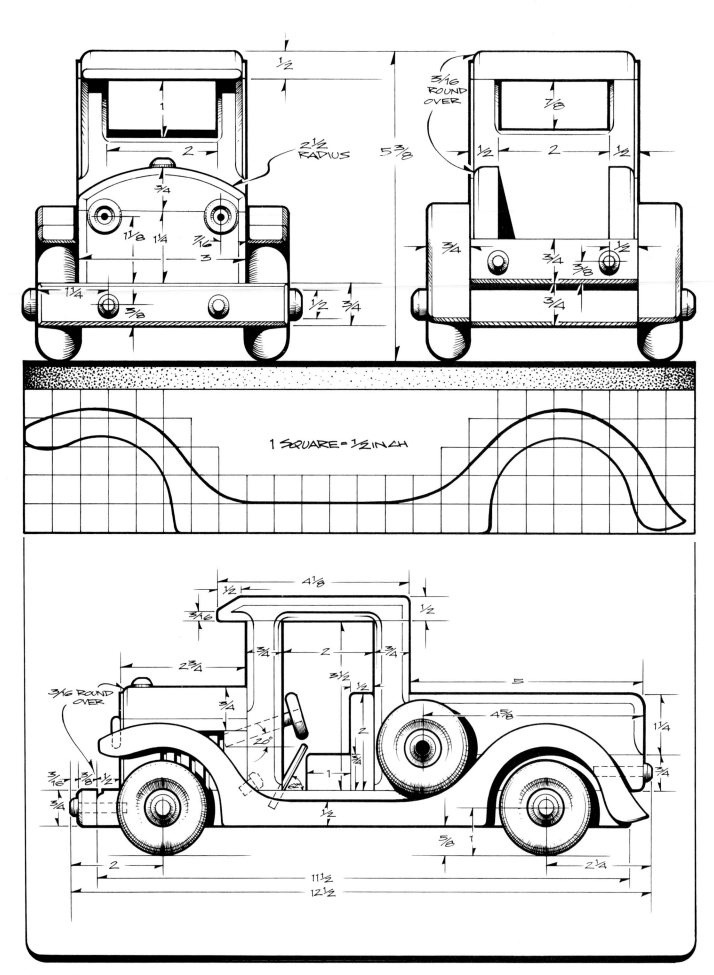

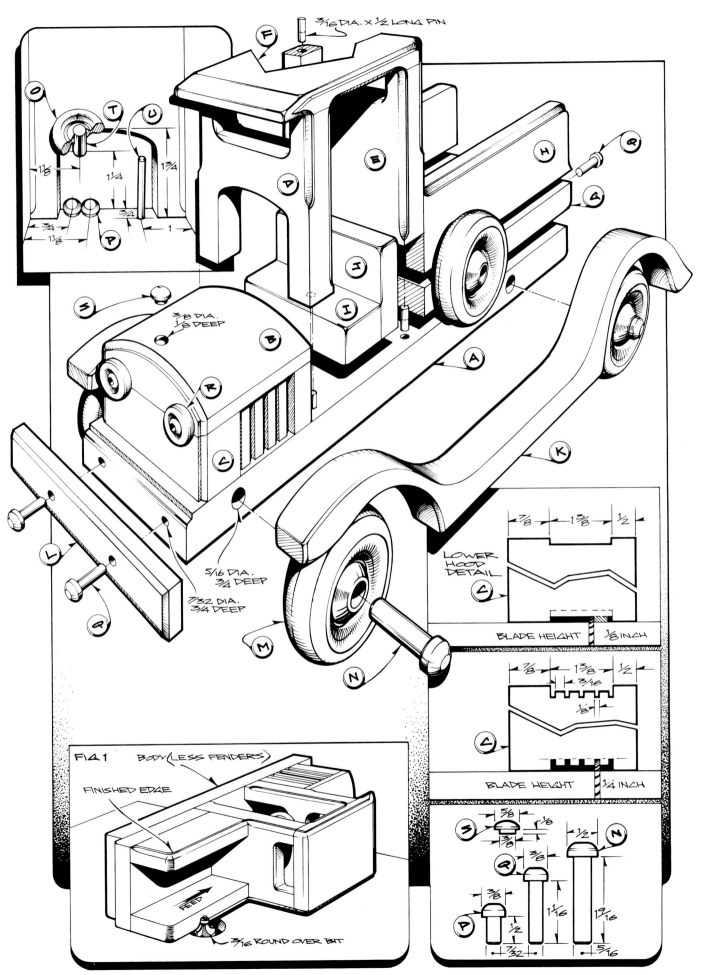

Whale Pull Toy

Our whale on wheels may just deliver your little Jonah from the oceans of plastic Rambos littering toy stores at Christmas time.

The birch and mahogany give the toy a rich texture and lasting appeal. Toymaker Skip Arthur designed this pull toy with an offset axle hole in the wheels so a toddler's eager pull opens the jaw as the body lurches along. The jaw opens because the wheels raise the body off the floor.

We show a band saw technique for making the whale, but you could just as easily use all hand tools. The power tools do save a lot of time, however, if you're going to turn out a school of whales for Christmas presents.

One tricky part of the construction is getting the tail to pivot freely on the body. A flathead screw holds the two parts together, with the shank serving as a bearing. It's tricky because the screw must be snug enough to hold the parts together, but loose enough to allow them to pivot freely. To get the right fit, cut a 1 in. by no. 10 screw down to ¾ in., and drill the holes in the tail and body as shown. You may need to adjust the screw length with a file.

If you're making the whale for a young child, omit the ball on the end of the string. It may present a choking hazard.

We prefer not to finish toys. If you choose to add one to the whale, make sure it's non-toxic.

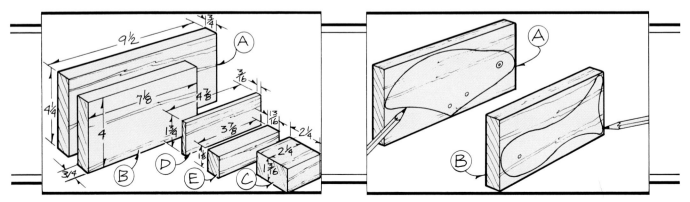

Step 1: Cut the seven blanks to size (there are two each of parts C and D). Parts A, B, C, and E are made from birch, parts D are mahogany.

Step 2: Using the full-size patterns provided, lay out the body (A) and tail (B) profiles and mark the location of the holes.

106   The Woodworker's Project Book

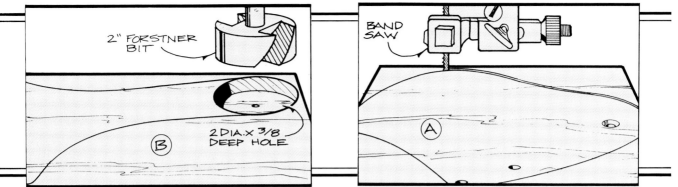

Step 3: Cut the half-lap pivot with a Forstner bit, drilling from one side on the body, and from the other side on the tail. Also drill for the axle, jaw pin and string hole, and the tail screw hole, which is countersunk.

Step 4: Use the band saw to cut the profiles on the body and tail. Soften the edges using a router with ¼ in. round-over bit. Use files where the stock is too thin to support the bearing of the router bit.

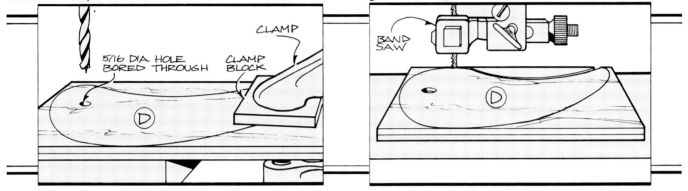

Step 5: Use the full-size pattern to lay out the profiles on parts D and E. Drill the 5/16 in. diameter jaw pin hole with parts D clamped together.

Step 6: Use the band saw to cut parts D and E. Use double-faced tape or a dab of hot glue to hold parts D together while cutting.

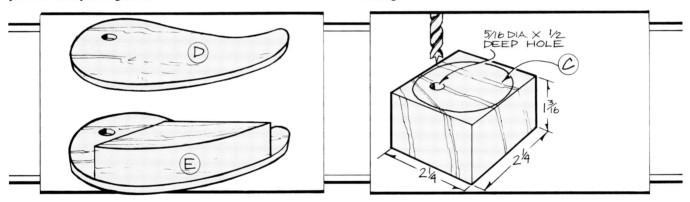

Step 7: Position parts D and E, then glue and clamp. The dotted line on the full-size pattern shows the position of E on D. Sand the edges flush if necessary.

Step 8: Lay out the wheels (C) on the blocks and drill the offset axle holes as shown in the full-size pattern.

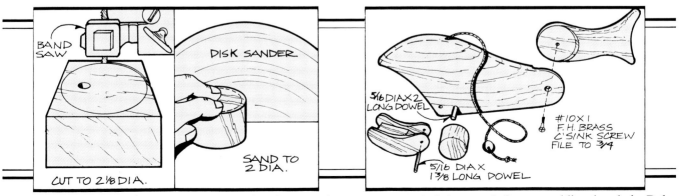

Step 9 (A and B): Cut out the wheels on a band saw and finish shape on a disk sander. Round over the edges. Also whittle a ½ in. diameter ball or disk from scrap, and drill a ⅛ in. diameter hole in it for the string.

Step 10: Sand all the parts before assembling the whale. Dabs of glue hold the dowels onto the wheels and jaw. But the dowels should pivot freely through the holes in the body.

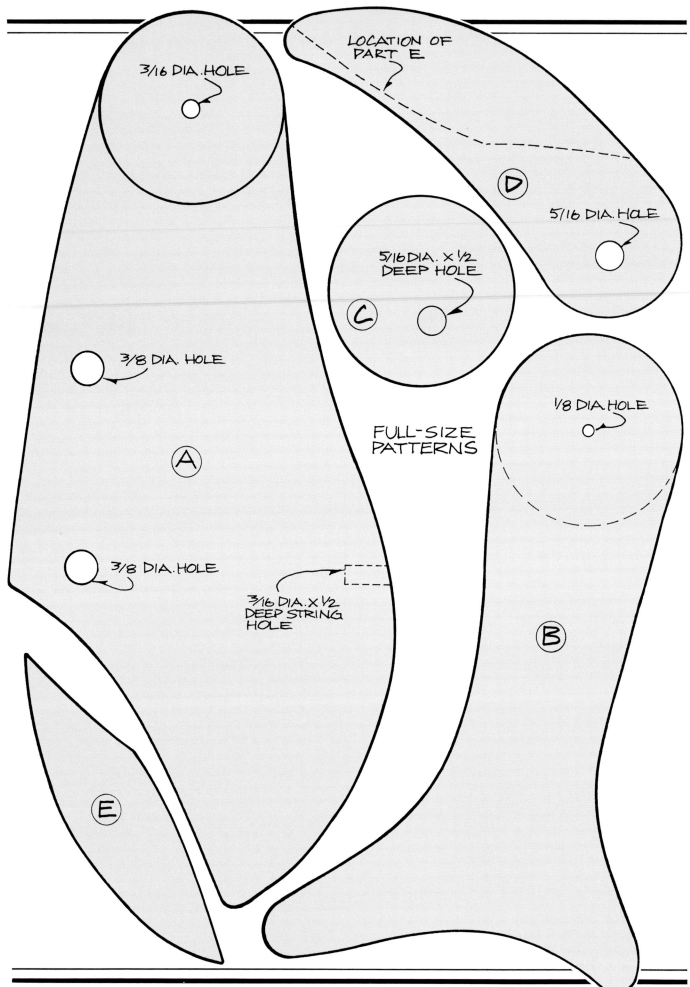

Child's Carousel Lamp

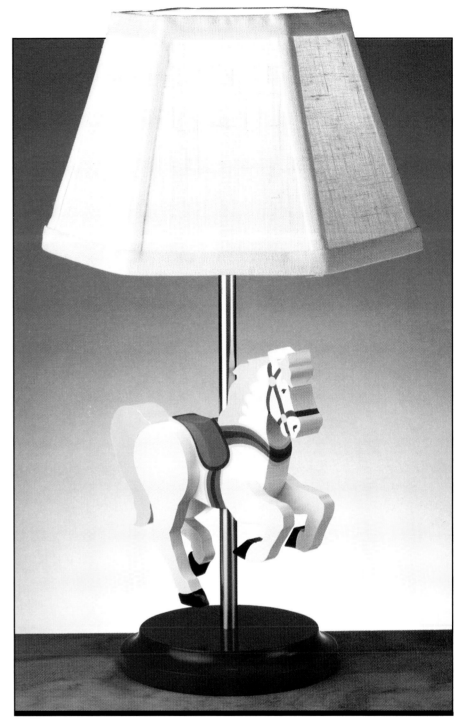

This lamp is the perfect gift for a child or grandchild. Fantasies of the gaily painted carousel pony will dance on long after the light has been turned off.

We used clear pine, 1 in. thick for the base (A) and body (B) and ½ in. thick for the legs (C, D, E). The 1 in. thick stock can be planed, and the ½ in. thick stock resawn from 1¼ in. stock, available at most lumberyards. The lamp hardware (parts F-M), and the shade (N) are available from a mail-order company (see Bill of Materials), if you can't find the parts locally.

If you can buy surfaced stock, a 1 in. thick board 6 in. wide by 14 in. long will yield the base and body, and a ½ in. thick board 6 in. wide by 10 in. long will yield all the leg parts. The two front legs are made from the same pattern. While you could trace the patterns, it's easier if you photocopy the full-size patterns, and paste them onto the stock with rubber cement or spray adhesive.

Cut the 1 in. thick stock into the sections for the base and body. Drill the hole through the body for the brass tubing (F), and drill the hole through the base for the cord (J). Also drill the holes in the base for the brass tubing and threaded lamp pipe (G), as well as the larger hole that allows access to the washers (K, L) and nut (M). Make the large diameter holes first, and then the ⅜ in. hole (See Base Detail). The scroll saw is used to cut out the round base and the body. Use the disk sander to clean up the base, and apply the cove detail with a ½ in. radius cove bit. A dowel wrapped with sandpaper can be used to sand the cove.

Sand thoroughly to remove the pattern paper and the rough edges from the scroll saw cuts. Once sanded, countersink the end of the cord hole in the base. Next, mark the locations of the legs on the body and label the legs so you don't get them confused. Once marked, they can be glued in place.

When the glue is dry, use spray enamel to paint the white on the horse, and the red on the base. Several light primer coats will make the finish more even. When dry, paint the details as shown in the full-size pattern. We used yellow for the mane and tail, red and yellow for the bridle, red, yellow and blue for the saddle, and black for the eyes, nostrils, and hooves.

Polish the brass tubing with 400-grit sandpaper, and coat it with clear lacquer or enamel. Be careful not to touch the brass tubing with your fingers after polishing it, or fingerprints will show under the lacquer. Then assemble the lamp as shown, roughing a small area on the tubing where it will be epoxied into the horse. Apply the epoxy to the tubing and slide the horse into position. Thread the cord through the base, hex nut (M), lock washer (L) and flat washer (K). Continue feeding it up the threaded rod (G) inside the tubing, then add the check ring (H) and the lamp socket (I). The threaded lamp pipe should be about 10⅞ in. long, but it's best to measure and cut exactly the length you need. File the ends after cutting to remove any burrs. Be sure to tighten the assembly before attaching the wire leads, so the cord will not twist.

(continued on next page)

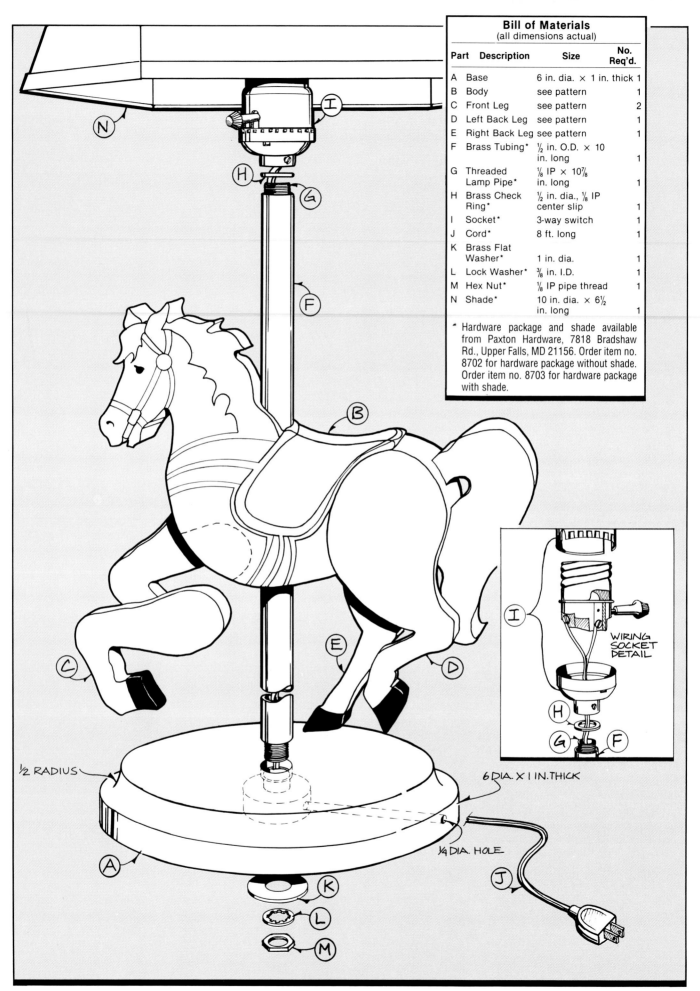

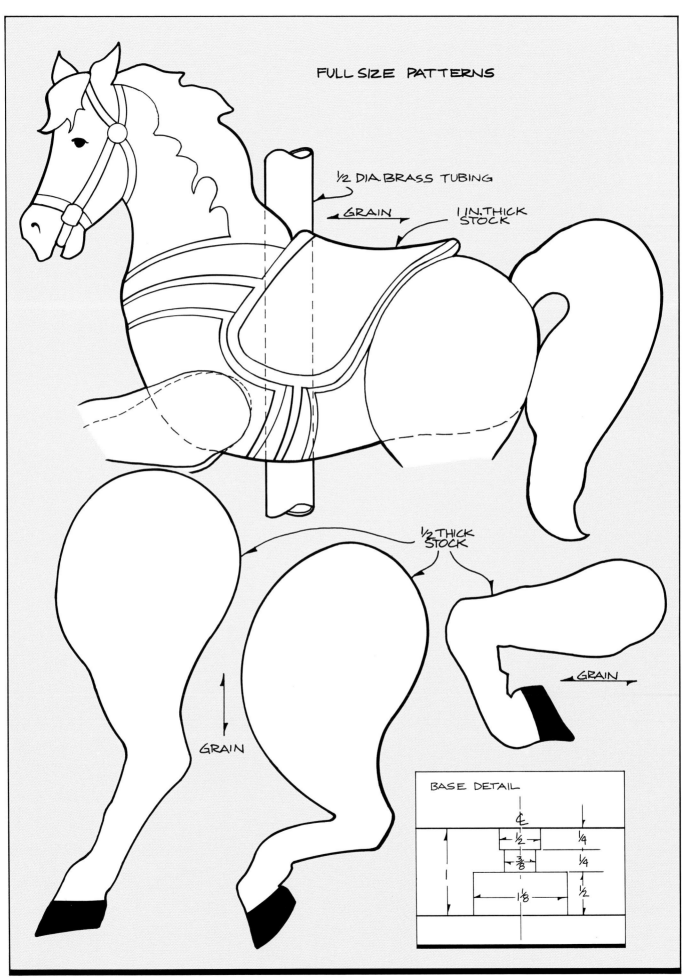

Colonial Wall Sconce

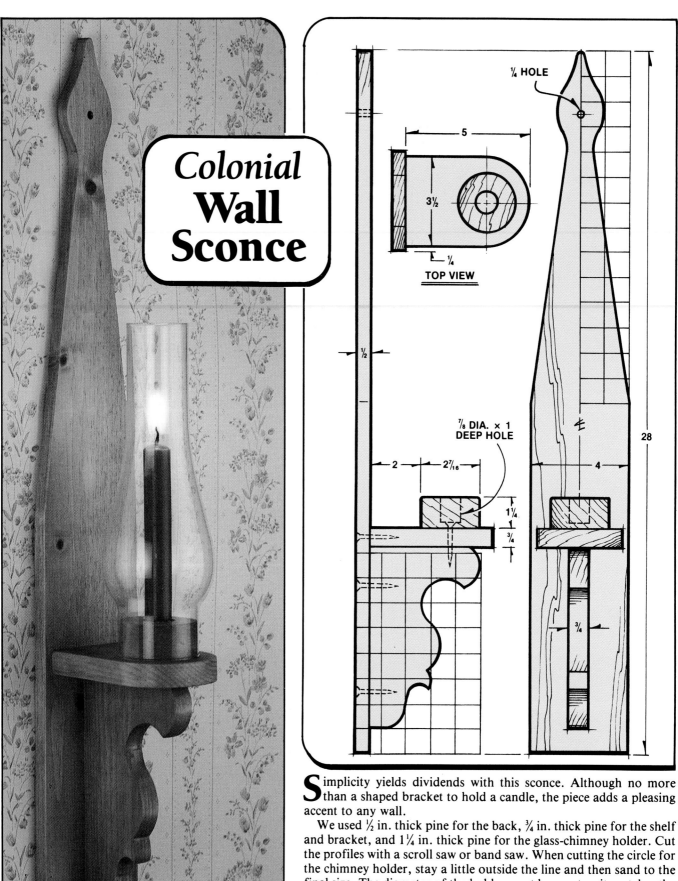

Simplicity yields dividends with this sconce. Although no more than a shaped bracket to hold a candle, the piece adds a pleasing accent to any wall.

We used ½ in. thick pine for the back, ¾ in. thick pine for the shelf and bracket, and 1¼ in. thick pine for the glass-chimney holder. Cut the profiles with a scroll saw or band saw. When cutting the circle for the chimney holder, stay a little outside the line and then sand to the final size. The diameter of the holder must be exact so it matches the opening of the chimney.

The four parts are screwed together as shown. For a finish, we used Minwax Ipswich Pine stain followed by two coats of tung oil. The 2½ in. by 10 in. glass chimney can be ordered from Paxton Hardware, 7818 Bradshaw Road, Upper Falls, MD 21156. Order part no. 8701.

Sources of Supply

The following pages list companies that specialize
in mail order sales of woodworking supplies.

United States

General Woodworking Suppliers

Constantine's
2050 Eastchester Rd.
Bronx, NY 10461

Craftsman Wood Service
1735 West Cortland Ct.
Addison, IL 60101

Frog Tool Co.
700 W. Jackson Blvd.
Chicago, IL 60606

Garrett Wade
161 Avenue of the Americas
New York, NY 10013

Highland Hardware
1045 N. Highland Ave., N.E.
Atlanta, GA 30306

Seven Corners Ace Hardware
216 West 7th Street
St. Paul, MN 55102

Shopsmith, Inc.
3931 Image Drive
Dayton, OH 45414-2591

Trend-Lines
375 Beacham St.
Chelsea, MA 02150-0999

Woodcraft Supply Corp.
210 Wood County Industrial Park
P.O. Box 1686
Parkersburg, WV 26102

Woodworker's Supply
5604 Alameda, N.E.
Albuquerque, NM 87113

W.S. Jenks and Son
1933 Montana Ave., N.E.
Washington, DC 20002

Hardware Suppliers

Anglo-American Brass Co.
Box 9487
San Jose, CA 95157-0792

Ball and Ball
463 West Lincoln Highway
Exton, PA 19341

Horton Brasses
P.O. Box 120
Cromwell, CT 06416

Imported European Hardware
4320 W. Bell Dr.
Las Vegas, NV 89118

Meisel Hardware Specialties
P.O. Box 70
Mound, MN 55364-0070

Paxton Hardware, Ltd.
P.O. Box 256
Upper Falls, MD 21156

Period Furniture Hardware Co.
Box 314, Charles Street Station
Boston, MA 02114

Stanley Hardware
195 Lake Street
New Britain, CT 06050

The Wise Co.
6503 St. Claude
Arabi, LA 70032

Hardwood Suppliers

American Woodcrafters
905 S. Roosevelt Ave.
Piqua, OH 45356

Arroyo Hardwoods
2585 Nina Street
Pasadena, CA 91107

Austin Hardwoods
2119 Goodrich
Austin, TX 78704

Bergers Hardwoods
Route 4, Box 195
Bedford, VA 24523

Berea Hardwoods Co.
125 Jacqueline Dr.
Berea, OH 44017

Maurice L. Condon
250 Ferris Ave.
White Plains, NY 10603

Craftwoods
10921-L York Rd.
Hunt Valley, MD 21030

Croffwood Mills
RD #1, Box 14J
Driftwood, PA 15832

Croy-Marietta Hardwoods, Inc.
121 Pike St., Box 643
Marietta, OH 45750

Dimension Hardwoods, Inc.
113 Canal Street
Shelton, CT 06484

Educational Lumber Co.
P.O. Box 5373
Asheville, NC 28813

Garreson Lumber
RD 3
Bath, NY 14810

General Woodcraft
531 Broad St.
New London, CT 06320

Hardwoods of Memphis
P.O. Box 12449
Memphis, TN 38182-0449

Henegan's Wood Shed
7760 Southern Blvd.
West Palm Beach, FL 33411

Kaymar Wood Products
4603 35th S.W.
Seattle, WA 98126

Kountry Kraft Hardwoods
R.R. No. 1
Lake City, IA 51449

Leonard Lumber Co.
P.O. Box 2396
Branford, CT 06405

McFeely's Hardwoods & Lumber
P.O. Box 3
712 12th St.
Lynchburg, VA 24505

Native American Hardwoods
Route 1
West Valley, NY 14171

Niagara Lumber
47 Elm Street
East Aurora, NY 14052

Sterling Hardwoods, Inc.
412 Pine St.
Burlington, VT 05401

Talarico Hardwoods
RD 3, Box 3268
Mohnton, PA 19540-9339

Woodcrafter's Supply
7703 Perry Highway (Rt. 19)
Pittsburgh, PA 15237

Wood World
1719 Chestnut
Glenview, IL 60025

Woodworker's Dream
P.O. Box 329
Nazareth, PA 18064

Wood Finishing Suppliers

Finishing Products and Supply Co.
4611 Macklind Ave.
St. Louis, MO 63109

Industrial Finishing Products
465 Logan St.
Brooklyn, NY 11208

The Wise Co.
6503 St. Claude
Arabie, LA 70032

Wood Finishing Supply Co.
100 Throop St.
Palmyra, NY 14522

WoodFinishing Enterprises
1729 N. 68th St.
Wauwatosa, WI 53212

Watco-Dennis Corp.
1433 Santa Monica Blvd.
Santa Monica, CA 90401

Clock Parts Suppliers

The American Clockmaker
P.O. Box 326
Clintonville, WI 54929

Armor Products
P.O. Box 445
East Northport, NY 11731

Klockit, Inc.
P.O. Box 542
Lake Geneva, WI 53147

Kuempel Chime
21195 Minnetonka Blvd.
Excelisor, MN 55331

S. LaRose
234 Commerce Place
Greensboro, NC 27420

Mason & Sullivan Co.
586 Higgins Crowell Rd.
West Yarmouth, MA 02673

Newport Enterprises
2313 West Burbank Blvd.
Burbank, CA 91506

Miscellaneous

Byrom International
(Router Bits)
P.O. Box 246
Chardon, OH 44024

Brown Wood Products
(Balls, Knobs, Shaker Pegs)
P.O. Box 8246
Northfield, IL 60093

Cherry Tree Toys
(Toy Parts)
P.O. Box 369
Belmont, OH 43718

Country Accents
(Pierced Tin)
P.O. Box 437
Montoursville, PA 17754

DML. Inc.
(Router Bits)
1350 S. 15th St.
Louisville, KY 40210

Floral Glass & Mirror
(Beveled Glass)
895 Motor Parkway
Hauppauge, NY 11788

Formica Corporation
(Plastic Laminate)
1 Stanford Rd.
Piscataway, NJ 08854

Freud
(Saw Blades)
218 Feld Ave.
High Point, NC 27264

Midwest Dowel Works
(Dowels, Plugs, Pegs)
4631 Hutchinson Road
Cincinnati, OH 45248

Homecraft Veneer
(Veneer)
901 West Way
Latrobe, PA 15650

MLCS
(Router Bits)
P.O. Box 4053
Rydal, PA 19046

The Old Fashioned Milk Paint Co.
(Milk Paint)
P.O. Box 222
Groton, MA 01450

Sears, Roebuck and Co.
(Misc. Tools & Supplies)
925 S. Homan Ave.
Chicago, IL 60607

Wilson Art
(Plastic Laminate)
600 General Bruce Drive
Temple, TX 76501

Canada

General Woodworking Suppliers

Ashman Technical
351 Nash Road North
Hamilton, ON L8H 7P4

Canadian Woodworker Ltd.
Unit 4 — 1391 St. James Street
Winnipeg, MB R3H 0Z1

House of Tools Ltd.
131-12th Ave. S.E.
Calgary, AB T2G 0Z9

J. Philip Humfrey International
3241 Kennedy Rd., Unit 7
Scarborough, ON M1V 2J9

Lee Valley Tools
1080 Morrison Dr.
Ottawa, ON K2H 8K7

Nautilus Arts & Crafts
6075 Kingston Road
West Hill, ON M1C 1K5

Sterling Tools
5043 Still Creek Ave.
Burnaby, BC

Stockade Woodworker's Supply
291 Woodlawn Rd. West, Unit 3C
Guelph, ON N1H 7L6

Tool Trend Ltd.
420 Millway Ave.
Concord, ON L4K 2V8

Treen Heritage Ltd.
P.O. Box 280
Merrickville, ON K0G 1N0

W. C. Robinson Woodworking Supplies
1615 Scugog Line, R.R. #1
Port Perry, ON L9L 1B2

Hardware Suppliers

Home Workshop Supplies
RR 2
Arthur, ON N0G 1A0

Lee Valley Tools
1080 Morrison Dr.
Ottawa, ON K2H 8K7

Pacific Brass Hardware
1414 Monterey Ave.
Victoria, BC V8S 4W1

Steve's Shop, Woodworking & Supplies
RR 3
Woodstock, ON M9V 5C3

Hardwood Suppliers

A & C Hutt Enterprises Ltd.
15861 32nd Ave.
Surrey, BC V4B 4Z5

Farrell Lumber Co.
1229 Advance Rd., Unit 3B
Burlington, ON L7M 1G7

Hurst Associates, Ltd.
74 Dynamic Drive, Unit 11
Scarborough, ON M1V 3X6

Longstock Lumber & Veneer
440 Phillip St., Unit 21
Waterloo, ON N2L 5R9

MacVeigh Hardwoods
339 Olivewood Rd.
Toronto, ON M8Z 2Z6

Unicorn Universal Woods Ltd.
4190 Steeles Ave. West, Unit 4
Woodbridge, ON L4L 3S8

Woodcraft Forest Products
1625 Sismet Road, Unit 25
Mississauga, ON L4W 1V6

Clock Parts Suppliers

Hurst Associates
105 Brisbane Road, Units 7–9
Downsview, ON M3J 2K6

Kidder Klock
39 Glen Cameron Rd., Unit 3
Thornhill, ON L3T 1P1

Murray Clock Craft Ltd.
510 McNicoll Ave.
Willowdale, ON M2H 2E1

Miscellaneous

Arbor Tools, Ltd.
(Tool Sharpening)
165 Limestone Crescent
Downsview, ON M3J 2R1

Bear Woods Supply Co.
(Pegs, Dowel Pins, Plugs, Toy Wheels)
Box 40
Bear River, NS B0S 1B0

Black & Decker
(Power Tools)
P.O. Box 9756
St. John, NB E21 4M9

Robert Bosch, Inc.
(Power Tools)
6811 Century Ave.
Mississauga, ON L5N 1R1

Freud
(Saw Blades)
100 Westmore Dr., Unit 10
Rexdale, ON M9V 5C3

Laurier Wood Craft
(Carving Supplies)
P.O. Box 428
South River, ON P0A 1X0

Nautilus Arts & Crafts
(Carving Supplies)
6075 Kingston Road
West Hill, ON M1C 1K5

R & D Bandsaws
(Custom made band saw blades)
41 Regan Road
Brampton, ON L7A 1B4

Also From Madrigal Publishing

The Woodworker's Journal

the magazine for woodworkers by woodworkers

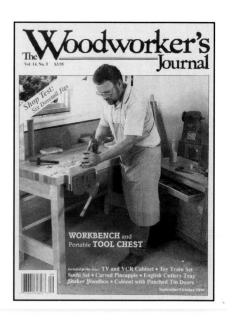

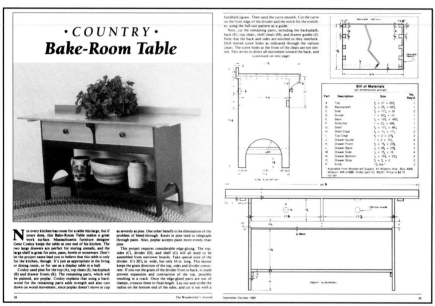

Over 50 great project plans each year!

Furniture ● Shop-Built Tools
Toys ● Easy Weekend Projects
Clocks ● Gifts

In-Depth Woodworking Articles!

- *Woodworking Basics* — fundamental skills for those just starting out
- *Special Techniques* — more advanced procedures
- *In The Shop* — safe use and care of tools
- *Finishing* — what you need to know about all kinds of finishes
- *Shop Tips* — quick ideas to save you time and trouble

The Practical Projects Magazine

It's like having a master woodworker at your side to help you plan, design, build and enjoy every project.

Subscribe Now!

$17.95 — One Year (6 issues)
$31.90 — Two Years (12 issues)

Madrigal Publishing Company
P.O. Box 1629, 517 Litchfield Road, New Milford, CT 06776
To order call toll-free 1-800-223-2425